D1565127

Sugar

A User's Guide to Sucrose

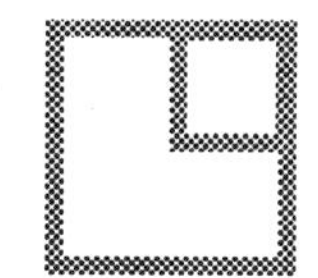

Sugar

A User's Guide to Sucrose

Edited by

Neil L. Pennington
Consulting Editor
The Sugar Association, Inc.
Washington, D.C.

and

Charles W. Baker
Vice President, Scientific Affairs
The Sugar Association, Inc.
Washington, D.C.

An AVI Book
Published by Van Nostrand Reinhold
New York

An AVI Book
(AVI is an imprint of Van Nostrand Reinhold.)

Library of Congress Catalog Card Number 89-21510
ISBN 0-442-00297-1

Manufactured in the United States of America

Published by Van Nostrand Reinhold
115 Fifth Avenue
New York, New York 10003

Chapman and Hall
2–6 Boundary Row
London, SE1 8HN

Thomas Nelson Australia
102 Dodds Street
South Melbourne 3205
Victoria, Australia

Nelson Canada
1120 Birchmount Road
Scarborough, Ontario M1K 5G4, Canada

16 15 14 13 12 11 10 9 8 7 6 5 4 3 2 1

Library of Congress Cataloging-in-Publication Data

Sugar, a user's guide to sucrose / the Sugar Association, Inc.
edited by Neil L. Pennington, Charles W. Baker.
p. cm.
Includes bibliographical references.
ISBN 0-442-00297-1
1. Sugar. 2. Sucrose. I. Sugar Association (U.S.)
II. Pennington, Neil L. III. Baker, Charles W.
TX560.S9S95 1990
641.3'36–dc20 89-21510
CIP

Contents

Preface ix

1 History of Cane and Beet Sugar 1

Sugarcane History 2
Sugar Beet History 5
The Sugar Industry Today 7
References 10

2 Processing Sugar From Sugarcane and Sugar Beets 11

Sugarcane Cultivation and Raw Sugar Processing 11
References 20
Cane Sugar Refining 22
Beet Sugar Production 26
References 35

3 Sugar Products 36

Granulated Sugar Products 36
Liquid Sugar Products 39
Brown Sugars 40
Specialty Sugars 41
Molasses 41
Conclusion 41
References 45

4 Properties of Sugar 46

Chemistry of Sucrose 46
Physical Properties of Sucrose 48
Chemical Properties of Sucrose 56
Biological Properties of Sucrose 57
Other Sugar Products 59
Storage and Handling of Sugar Products 60

Other Sweeteners 63

5 The Flavor of Sugar in Foods 66
Flavor of Sucrose 66
Other Sugar Products 69
References 70

6 The Sweetness of Sugar 71
Measurement 71
Validity of Measures 73
Differences Among Sweeteners 74
Interactions Between Sweet Taste and Other Sensations 74
Differences Among People 76
Pleasantness of Sucrose 76
References 78

7 Sugar in the Body 82
Production in Green Plants 82
Utilization of Sugar in Humans 85
Sugar and Health 88
References 100

8 Sugar in Confectionery 103
Sugar Properties Important to Confectionery 103
Basic Candy Types 110
References 128

9 Sugar in Bakery Foods 130
Sucrose Products in Baking 130
Yeast-Leavened Bakery Foods 132
Chemically Leavened Bakery Foods 139
Icings and Fillings 145
Summary 150
References 150

10 Sugar in Dairy Products 152
Sugar in Frozen Dairy Products 152
Chocolate-Flavored Milk Products 157
Sweetened Condensed Milk 158
Egg Nog Products 159
Sugar as Part of Flavorings for Dairy Products 160

Summary 163
References 164

11 Sugar in Processed Foods 165

Caramelization and Maillard Browning Reactions 166
Water Activity and Sugars in Food Manufacture 168
Agglomeration Processes and Chocolate Drink Products 169
Sugars in Fruit Processing 171
Tomato Catsup, Chili Sauce, and Barbecue Sauce 171
Fruit-Flavored Beverage Powders 172
Citrus Purée Base for Citrus Beverages 174
Lemon Pie and Chiffon Fillings 174
Instant Creamy "No-Bake" Pie Fillings 175
Gelatin Dessert Mix 176
Lemon-Flavored Iced Tea Mix 177
Processed Meats 177
References 181

12 Sugar in Ready-to-Eat Breakfast Cereals 182

The Ready-to-Eat Breakfast Cereal Industry Today 183
Breakfast Cereal Manufacturing Processes 184
Sugar and R.T.E. Cereals 187
Sucrose Content of R.T.E. Cereals 192
Sugar by the Spoonful 195
Conclusion 195
References 195

13 Sugar in Beverages 198

Carbonated Beverages 198
Powdered Drink Mixes 210
Alcoholic Beverages 210
References 211

14 Sugar in Preserves and Jellies 212

Today's Products 214
Scientific Principles 217
Quality Attributes 219
Modern Manufacturing 221
Typical Formulas 222
References 226

15 Sugar in Microwave Cooking **228**

Microwave Energy 230
The Food Materials 235
Factors Affecting Microwave Cooking 248
Special Role of Sugar 256
Major Applications 256
Selected Recipes 257
Appendices 263
References 273

16 Nonfood Uses for Sucrose **276**

Sucrochemistry 278
Fermentation Products 281
Pharmaceutical Applications 283
Cosmetics Applications 284
Other Applications 285
Conclusion 285
References 286

17 Methods of Analysis of White Sugar **288**

18 Sugar Industry Terminology **303**

19 The Sugar Industry **312**

A. Refined Sugar Producers 312
B. Cane Raw Sugar Producers—USA 314
C. Sugar Industry Affiliates 316

Index **319**

Preface

Sucrose is a basic carbohydrate and has occupied a central position in human food for centuries. Sugar contributes to the pleasant taste and physical structure of many foods. This monograph is intended to provide an overview of the numerous roles of sucrose in various food and nonfood processes. *Sugar: A User's Guide to Sucrose* has been structured as a basic reference source for the fundamentals of sucrose functionality and utility.

Authors from the food industry and academia were selected on the basis of their recognized expertise in the functional roles of sucrose in the formulation and final quality of a specific food. Each author writing about a specific food has described why sucrose-based formulations are the standards against which alternative formulations or processes are judged. In addition to the fundamentals of sucrose usage in foods, other authors have communicated the role of sucrose in flavor and taste, a balanced diet, microwave cooking, and the synthesis of specialty chemicals such as detergents, pharmaceuticals, and cosmetics.

The histories of cane and beet sugar development and processing were included to show that sucrose has been an important agricultural commodity and food for centuries. Although the number of specific sugar products is more numerous in today's world, refined sucrose is essentially the same today as in past centuries. From its formation during photosynthesis through its refinement to a high-quality product, sugar is an essential ingredient in a wide range of foods and an important source of food energy.

Scientists and technologists in the food and beverage industries will be interested in the discussion of the physical and chemical properties of sucrose that impact its processing, food-specific functionality, and behavior under the influence of microwave heating and cooking, and also the analytical methods used to ensure sucrose quality. These sections will also satisfy the requirements of educational or apprentice programs of the food and beverage industries. Terminology common to the sugar industry and the listing of sugar industry companies, associations, and affiliates were compiled for the benefit of various trade and professional associations, production and marketing management, and food media professionals.

Sugar: A User's Guide to Sucrose is the result of the efforts of eighteen authors. The co-editors thank each author.

Sugar

A User's Guide to Sucrose

1

History of Cane and Beet Sugar

Laszlo Toth, Ph.D., and A. B. Rizzuto*

History is defined as a chronological account of what has happened of significance in the life and development of a people, country, or institution; also it can be a scientific account of a system of natural phenomena. A thorough history of sugar would be very lengthy because it began millions of years ago. The creation of sugar, which the chemist calls sucrose, belongs to the infinite wisdom of nature. The sophisticated chain of reactions that first occurred millions of years ago still goes on in the majestic laboratories of Mother Nature where solar energy is captured in certain plants to form sugar. Its sweet flavor resulted from pure solar energy centuries ago as it does today, providing energy required by each living being to sustain life.

Rather than starting at the beginning, this chapter will address events related to the development of the industry that today processes and packages sugar. Statistics record that the world today manufactures and consumes over 110,000,000 tons of sugar per year. Taking into consideration that primitive domestic sugar manufacturing, mostly in South and Southeast Asia, is not included in this figure, total yearly world sugar production could be close to 125 to 130 million tons. This tremendous amount of product is an important, worldwide source of human energy.

The industrial production of sugar/sucrose is based exclusively on sugarcane and sugar beet processings. Sugarcane is a tropical plant, while the sugar beet flourishes in cooler climates. Sugarcane is grown and cane sugar is produced in tropical and subtropical countries. All these regions are inside a belt, whose northern

*Dr. Laszlo Toth is Technical Director, Western Sugar Company, Denver, Colorado. A. B. Rizzuto is Director of Research, Amstar Sugar Corporation, Brooklyn, NY.

borderline crosses the North American continent at southern California and South Carolina, the European continent at the southern tip of the Iberian Peninsula and Sicily, the Asian continent at the Middle East countries through Pakistan and south China, and the Pacific Ocean at the 37th parallel. In the Southern Hemisphere the borderline goes through the southern tip of Brazil, crosses the Atlantic Ocean, the African continent at Natal, the Indian Ocean, the northeastern coast of Australia, and the Pacific Ocean on the 34th parallel. In all regions outside this belt, in the northern and southern hemispheres, the sugar beet dominates.

Despite the fact that refined cane and beet sugar are physically as well as compositionally identical materials, the colorful histories of their development are very different.

In the very beginning sugar was not known in its solid form but only as a sweet syrup obtained from sugarcane. Its name originated from the Sanskrit word "Sharkara," which denotes "material in a granular form," and is the origin of the term "sugar" in modern languages—Arabic "suchar"; French "sucre"; German "zucher"; Hungarian "cukor"; Russian "sahar"; Spanish "azuckar." The story of both sugars, cane and beet, is woven into historic tales of adventure and discovery. It is important in trade and commerce today as in the past. Sugar has played a fascinating role in the destinies of nations in war and peace.

Sugarcane History

The oldest source for sugar manufacturing is sugarcane. There are indications of primitive sugar manufacturing in New Guinea more than 12,000 years ago.

Sugarcane production spread southeastward to the New Hebrides and New Caledonia in about 8000 B.C. In approximately 6000 B.C., sugar manufacturing appeared in Celebes, Borneo, Java, Indochina, and India. From India, sugarcane was carried to China, where soil conditions were ideal for its growth. Chinese literature records that in about 200 B.C. the Emperor Tai-Sun sent his emissary to India to learn cane sugar manufacturing. Also, the Chinese emperor received sugarcane as a tribute from the kingdom of Funam. Most likely the sugarcane arrived in Persia, Arabia, and Egypt about the same time. In the meantime, it also found its way into Syria, Palestine, and the island of Cyprus.

The first written records in European literature dealing with cane sugar date from the era of Alexander the Great's war with India, 327 B.C. His army commanders, Nearhos and Onescritos, reported on "honey that is produced from cane, without participation of honey bees." Cane sugar's trek to numerous northeastern Pacific islands and Hawaii took place in quite recent history, mostly during the first millennium A.D.

Closer knowledge in Europe about sugar was spread during the Crusades in the eleventh to thirteenth centuries. Many people from various European nations and social levels fought in this series of campaigns. Some records show that large cane plantations belonged to various Christian orders in the region. Shipping and trade between the Mideast and Europe during this period was almost exclusively in the hands of Venetian merchants. There are indications that, besides raw sugar trade, the Venetians also invented some primitive sugar refining. There is evidence that in the middle of the fourteenth century the Venetians were manufacturing and trading "sugar loaves" with their customers.

The various Arab armies brought sugarcane with them during their conquests across the northern coast of Africa. Possibly, from there it made its way into Spain and Sicily about 700 A.D., though there are indications that some sugarcane might have been in the Mediterranean coastal lands earlier. According to some sources, the Moorish conquering army brought sugarcane plants along with their science and inventions.

Sugar was one of the first pharmaceutical ingredients used, as it still is today, to mask the bitter or unpleasant taste of medicines. Sugar was so rare that a teaspoon of it in the sixteenth century commanded the equivalent of five dollars. According to English records from the seventeenth century, one could purchase a calf for four pounds of sugar. The price was still very high in the eighteenth century.

The quest for sugar and spices ushered in the age of discovery. The Portuguese were leading explorers and colonizers. They established sugar as a crop in Madeira, the Azores, and the Cape Verde Islands. The Spanish brought cane to the Canary Islands. Slave labor in these colonies produced sugar at a lower cost than possible in Mediterranean countries.

On his second voyage to the New World, Columbus brought cane cuttings to Hispaniola, the island we know now as the Dominican Republic and the Republic of Haiti. The first attempt to grow cane failed, but by 1509 sugar was being produced in profitable amounts.

Sugar shaped a good deal of the history of the New World. By 1520 cane was growing in Mexico, and the Spanish explorer Cortez established the first North American sugar mill there in 1535. Cultivation soon spread to Peru, Brazil, Columbia, and Venezuela. Puerto Rico had a mill by 1548. By 1590 more than one hundred sugar mills were flourishing in Brazil alone. In 1624 Brazil fell to the Dutch, but Dutch rule lasted only until 1654, when the Portuguese took over again and expelled 20,000 Dutch. Many of them migrated to the West Indies, where they contributed their knowledge to the production of sugar.

The "sugar islands" of the West Indies brought great wealth to England and France. Queen Elizabeth displayed her wealth by putting a sugar bowl on her table and using sugar as an everyday food and seasoning. It soon appeared at court banquets. Great Britain took a commanding position in the sugar trade, and it was not long before the introduction of coffee, chocolate, and tea in the English diet tremendously increased the use of sugar. By the end of the seventeenth century, these new beverages were in general use, and by the eighteenth century the demand for sugar was so great that it became a matter of active public interest. Historic records show that large amounts of revenue for the Spanish and British crowns were derived from the great sugar plantations along the Spanish mainland and from the British islands of Jamaica, St. Kitts, and Barbados.

The Jesuits, who introduced cane cultivation to the Argentines in 1670, were the first to bring sugarcane to what is now the United States. In 1751, the first crop was planted in Louisiana with cuttings brought from Santo Domingo. The cane thrived, but no progress was made in extracting sugar. Planting on a large scale was abandoned in 1776, and the industry languished until 1791, when it was again tried, this time with success.

On the other side of the world, the island of Mauritius in the Indian Ocean started production in 1747. In 1824, cane growing was introduced in Australia.

In other places later to become part of the United States, sugarcane cultivation was slower to take hold. Sugarcane is reported to have been grown in New Smyrna, Florida, as early as 1767; however, it was not until the early 1920s that the first extensive commercial plantings were established near Canal Point east of Lake Okeechobee (7). Sugarcane was grown for home use in the Hawaiian Islands prior to 1778, but the first large-scale plantings were not started until 1825 (8). The initial plantings of sugarcane in Texas were early in the nineteenth century. However, it was not until 1913, in the Rio Grande Valley, that the Texas sugarcane industry reached a high of five sugar mills in operation (9). This peak in the Texas sugar industry was short-lived due to economic and political conditions which caused the last mill to close in 1921. The present Texas sugar industry was established in the Rio Grande Valley when a group of one hundred farmers formed a sugarcane cooperative in 1970.

Sugar Beet History

Sugar beet cultivation and the extraction of sugar from the plant is a younger technological development than that of sugarcane. Since ancient times, a white-colored beet called the "Beta Maritima," has been grown in Mediterranean countries, but it was not until 1590 that anyone recorded its sweet properties. That year, Oliver De Serres prepared a syrup from beets and noted that the "juice yielded on boiling is similar to sugar syrup." More than 150 years later, in 1747, Andreas Marggraf, a member of the Berlin Academy of Sciences, became interested in the "sweet root's" properties. During his laboratory tests he succeeded in extracting sugar from thin slices of beets, using alcohol, and in crystallizing it. No practical use was made of this discovery until 1798, when F. C. Achard, a German pharmacist who had been a student of Marggraf, separated ten pounds of sugar in a small pilot plant. He presented this refined beet sugar and his written report to the King of Prussia. The report described methods used and the economic importance of developing a new industry based on extracting and crystallizing sugar from sweet roots of "Beta Maritima." This resulted in the establishment of the first beet sugar factory on the estates of Kunern in Silesia. The first processing started in April 1802, and with relatively primitive technology coupled with low beet sugar content, the factory was able to produce about three pounds of sugar for each hundred pounds of beet. Despite the very low efficiencies and high sugar losses, this factory and other early beet sugar factories were successful due to the high sugar prices prevailing at that time.

The real push for the development of a large-scale beet sugar industry came with the Napoleonic Wars and the continental blockade in 1806. The blockade prevented any merchandise from England or its colonies from entering the European continent. Since the only sugar available to most European countries at that time was shipped overseas from cane plantations in their tropical colonies, a severe sugar shortage occurred. To overcome this shortage,

Napoleon's French government offered a bounty on beet sugar, established six special beet sugar schools for select students, set aside large tracts of land, and compelled the peasant farmers to plant sugar beets. These various governmental acts, together with the high price of sugar, resulted in rapid development of the new technology. In 1813, only eleven years after the start-up of the first beet sugar factory, and only seven years after the continental blockade, 334 beet sugar mills were operational in France, Germany, and Austria-Hungary. The factories of the time were of very small capacities, barely processing one hundred tons of beets per day, with primitive equipment and poor quality beets that usually contained no more than 5 to 6 percent sugar.

After the Napoleonic Wars, the cane sugar supply started its free flow again, causing a serious but not long-lasting slowdown in the young beet sugar industry. Progress had been made in developing better-quality beets. The process of beet selection had already been started by Achard, but in 1821 Pelouze, through research in plant breeding, was able to produce a dramatic increase in the sugar content of beets. This improvement, together with high sugar prices, allowed the industry to progress rapidly again, so that by 1836 there were 436 beet sugar factories in operation.

The young beet sugar industry received a new shot in the arm with the revolt of slave labor on the sugarcane plantations of Santo Domingo. The interruptions in overseas cane sugar production caused a further strain on the sugar supply, and prices increased consequently, encouraging further beet sugar development.

In the second half of the nineteenth century, a series of improvements, inventions, and new ideas caused radical changes in beet sugar technology and led to more efficient and economical processing. These processes are still in use today. In 1840, lime-carbon dioxide juice purification was introduced. In 1843, the multiple-effect evaporator, developed for the cane sugar industry, was put in service. In 1844, centrifugal separation of crystals from *massecuite*, a dense syrup containing sugar crystals, was begun. In 1863, the Roberts diffusion battery was employed to extract sugar from beets, and by 1870, the vacuum pan and efficient crystallization technology were coming into general use. Patented in 1883, the Steffen Process increased sugar recoveries. In 1884, the idea of crystallization dynamics was proposed and resolved by G. Z. Wulff. All these developments created a strong and vital industry. By 1914 world production of beet sugar was 9,051,767 tons, compared with cane sugar production of 11,523,158 tons.

The development of the beet sugar industry in the United States paralleled that of Europe. Many failures were experienced during the early years of the industry. In 1838, the first factory in the United States was erected in Northampton, Massachusetts, by David Lee Child. It was a failure and ceased to operate after 1840. Between 1838 and 1870 there were several unsuccessful attempts to erect and run a beet sugar factory: White Pigeon, Michigan, in 1848; Salt Lake City, Utah, built by the Mormons in 1853; a factory built by the Gennett brothers in Chatsworth, Illinois, in 1866; a factory built by Otto and Bonesteel in Fond du Lac, Wisconsin, in 1868; and a beet sugar plant built by E. H. Dyer in Alvarado, California, in 1870. In 1879, E. H. Dyer, after four complete financial failures and reorganizations, succeeded in putting the Alvarado plant on a paying basis. In 1888, Claus Spreckels started a successful plant in Watsonville, California.

From that time on, the growth of beet sugar in the United States has been fairly constant. In 1915, seventy-nine beet sugar factories were in operation, and by 1919 the number had grown to one hundred.

The Sugar Industry Today

The American cane and beet sugar industries of 1990 reflect the dramatic changes caused by the development in the early 1970s of isomerized corn syrups and nonnutritive, artificial sweeteners. Though cane and beet sugars remain the standard measure of quality, taste, and functionality, corn syrups and artificial sweeteners have each established themselves as substitutes for sugar in segments of the sweetener market. Figure 1-1 shows U.S. per capita sweetener consumption from 1980 to 1989.

Despite competition from new products, the American cane and beet sugar industry remains vital, as the latest statistical data show. Present U.S. sugar consumption is approximately 8.35 million tons per year, and according to forecasts, it will maintain the same general level for the next few years. Cane sugar products are processed in twelve refineries with production of approximately five million tons per year of sugar. Thirty-eight sugar beet factories produce about 3.5 million tons per year. In order to supply the cane sugar refineries, about eight hundred thousand acres of cane are harvested in Florida, Louisiana, Texas, and Hawaii and milled in forty-two raw sugar plants. An additional one million tons of raw sugar are imported each year. Beets are harvested on 1.3 million acres of land and processed in thirty-eight sugar factories in thirteen states. Figures 1-2 and 1-3 compare U.S. and world sugar production yields of sugarcane and sugar beet.

The contemporary sugar industry represents a vital part of the U.S. economy and employs a large number of people in such activities as the growing and harvesting of beet and cane sugar, cane refining, raw sugar milling, cane and beet sugar processing, research, sales, and service. In addition, an enormous number of suppliers of raw materials, chemicals, machines, and parts amount to several hundred additional people for each factory. Most important are the countless industrial users of sugar that depend on a steady, reliable source and the grocery stores that supply the consuming public. In the U.S.A. alone, the sweetener industry directly and indirectly generates $18.5 billion per year in wages and revenues and give jobs to 361,000 employees in forty-two states.

So, sugar processing has come a long way in the 12,000 years since its beginnings in New Guinea.

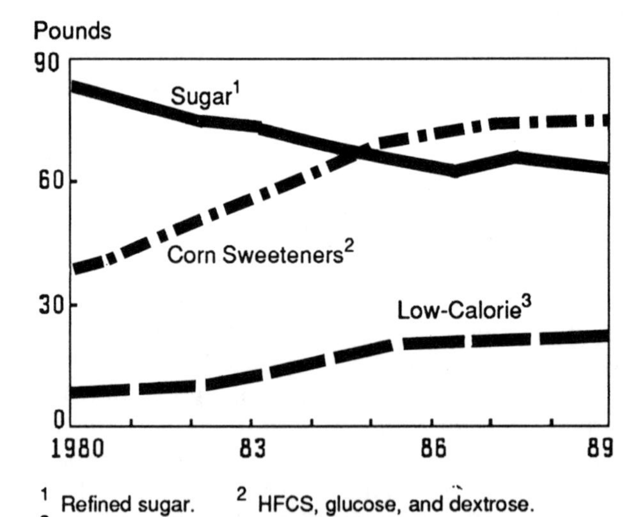

Figure 1-1. U.S. per capita sweetener consumption, 1980-89. (Source: U.S. Department of Agriculture, *Sugar and Sweetener Situation and Outlook Report*, Dec. 1989.)

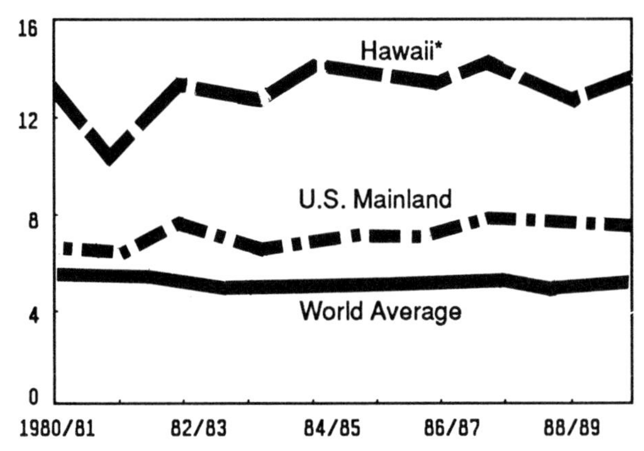

Figure 1-2. U.S. and world cane sugar yields, 1980-89. (Source: U.S. Department of Agriculture, *Sugar and Sweetener Situation and Outlook Report*, Dec. 1989.)

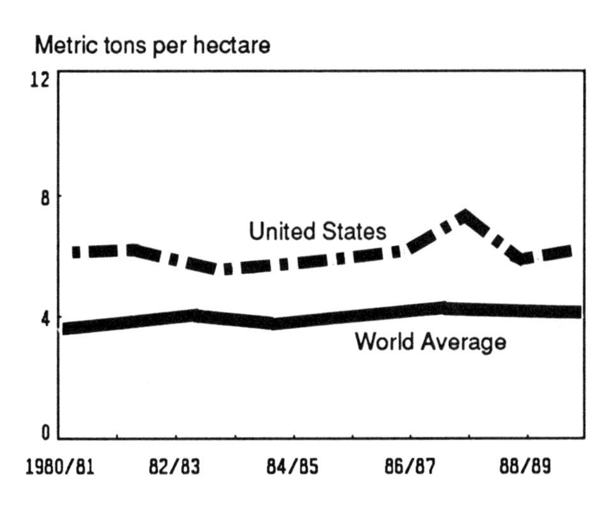

Figure 1-3. U.S. and world beet sugar yields, 1980-89. (Source: U.S. Department of Agriculture, *Sugar and Sweetener Situation and Outlook Report*, Dec. 1989.)

References

1. Baikov, V. E. *Manufacturing and Refining of Raw Cane Sugar*. Elsevier, Amsterdam, London, New York. 1967.
2. Jenkins, G. H. *Introduction to Cane Sugar Technology*. Elsevier, Amsterdam, London, New York. 1966.
3. Paladino, B. *Technologija Secera*. Industriska Knjiga, Belgrade. 1948.
4. Susic/Guralj. *Osnovi Tech. Sec.* Naucna Anjiga, Belgrade. 1965.
5. Schneider, F. *Tech. Des Zuckers*. Verlag M. H. Schaper, Hanover. 1968.
6. McGinnis, R. A. *Beet Sugar Technology*. Beet Sugar Development Foundation, Ft. Collins, CO. 1982.
7. Sitterson, C. J. *Sugar Country*. Kentucky Press, Lexington, KY. 1953.
8. Deerr, N. *The History of Sugar*. Vol. I. Chapman and Hall, London. 1949.
9. Rio Grande Valley Sugar Growers, Inc. *The Texas Sugar Cane Industry*. November, 1989.

2

Processing Sugar From Sugarcane and Sugar Beets

Henry J. Andreis, A. B. Rizzuto, and S. E. Bichsel

This chapter is divided into three sections: the first on the cultivation of sugarcane and the processing of it into raw cane sugar; the second on the production of refined sugar from raw cane sugar; and the third on the production of refined sugar from sugar beets.

Sugarcane Cultivation and Raw Sugar Processing

Henry J. Andreis*

Cultivation of Sugarcane

Sugarcane was introduced into the New World by Christopher Columbus and grown successfully there from the sixteenth century on; but, as described in Chapter 1, it was not successfully grown commercially in the United States until much later (1,2). Today sugarcane is grown commercially in Florida, Hawaii, Louisiana, and Texas. Based on 1988 production records, Florida produced 1,595,000 tons of raw sugar, followed by Hawaii with 929,000, Louisiana with 797,000, and Texas with 108,000 (3).

Commercial Growing Practices. Sugar production varies widely from field to field and between sugar-growing areas, ranging from below two to over eight tons per acre per year. This variation in sugar

*Henry J. Andreis is Vice President and Director of Research, United States Sugar Corporation, Clewiston, FL.

production is influenced by many factors, including length of growing season, soil quality, soil fertility, varieties grown, weather conditions, diseases, and insects.

Sugarcane is primarily a tropical plant that takes from eight to over twenty-four months to ripen. Sugars synthesized in the leaves are used as a source of energy for growth or are translocated to the stalks for storage. It is the stalks that are harvested commercially for cane sugar production.

In farming operations, sugarcane is reproduced vegetatively by planting either whole stalks or pieces of stalks 18 to 36 inches in length (seedpieces). Seedpieces contain one bud at each node, and when planted, these buds germinate and develop into new plants with the same characteristics as those found in the original clone. While sugarcane often flowers and produces true seed, these seeds are used only in breeding programs to develop new varieties because each germinated seed produces a plant that is genetically different from all others.

Seedcane is cut to appropriate lengths either by hand or by machine prior to being delivered to the fields for planting. These pieces are dropped into newly opened furrows by workers walking behind trailers or by mechanical planters (Figure 2-1). Three to over five tons of seedcane are needed to plant one acre; the higher rate is needed when mechanical planters are used. If required, fertilizer and insecticides are placed in the furrows as they are formed or when they are closed. Depending on location, the seedpieces are covered with 2 to 8 inches of soil. In tropical locations, where rapid germination is desirable, the seedpieces are covered with only 2 to 4 inches of soil. However, in subtropical areas, 6 to 8 inches of soil are used to protect the seedpieces from injury by the cold. Herbicides and mechanical cultivation are necessary to control weeds until the sugarcane plants become large enough to shade the ground. Once the cane closes in, no equipment enters a sugarcane field until harvest.

Several successive crops may be harvested from a cane planting. The first crop after planting sugarcane is known as plant cane, while the following crops are identified as first ratoon, second ratoon, third ratoon, and so on. Fertilizer is applied to stubble crops by banding next to the rows of cane, by broadcasting or, where drip irrigation systems are used, by adding it in the irrigation water. As in the plant cane crop, stubble crop weeds and insects are controlled by cultivation and chemicals as needed.

Figure 2-1. Workers planting sugarcane by dropping pieces of cane into furrows. (Source: Kim R. Sargent, U.S. Sugar Corp.)

Sugarcane grown in Florida, Texas, and dry areas of Hawaii is irrigated, while that grown in Louisiana and high-rainfall areas of Hawaii is nonirrigated. Types of irrigation used include drip, open ditch, furrow, and overhead.

When sugarcane matures, harvest operations commence (Figure 2-2). Where possible, fields of sugarcane are tested for sucrose content, and the most mature fields are harvested first. In Florida, Hawaii, and Texas, standing cane is fired to burn off the dry leaves, while in Louisiana the cane is cut and laid on the ground before burning. The burning speeds up harvesting, reduces hauling costs, and supplies the mills with cleaner cane. Cane is machine harvested in Hawaii, Louisiana, and Texas, and both hand and machine harvested in Florida. The harvested cane stalks are loaded mechanically into trucks or railroad cars and hauled to mills for processing into raw sugar (Figure 2-3).

Commonly, after two to three crops of sugarcane are harvested from a field, production becomes too low to be profitable, and the field is plowed up and prepared for a new planting. While a high percentage of fields are left fallow for a few weeks to several months prior to replanting sugarcane, some are planted with a cover crop such as corn, rice, cotton, winter wheat, or soybeans. The rice, corn,

Figure 2-2. Sugarcane harvester cutting and loading cane to be transported to trucks or railcars. (Source: Henry J. Andreis, U.S. Sugar Corp.)

cotton, and winter wheat are generally harvested, while the soybeans are usually plowed under for use as a "green manure."

Sugarcane Varieties. Sugarcane varieties developed at breeding stations located in Florida, Hawaii, and Louisiana and evaluated in the variety-testing programs in all four sugarcane states are essential to maintaining a viable sugarcane industry. Sugarcane breeders develop varieties that can be grown profitably under the varied U.S. environmental conditions. Since sugarcane is a tropical plant, it takes strong breeding efforts to develop varieties that ripen during the eight-month growing season of Louisiana. Because there are no freezes in Hawaii, varieties adapted to a twenty-four-month growing season have been developed to reduce production costs. In Florida, breeders have developed varieties that have the ability to produce and store sugar when grown on nitrogen-rich organic soils.

Sugarcane varieties grown in Florida are obtained from either the USDA-ARS Sugarcane Field Station at Canal Point or the United States Sugar Corporation Research Station at Clewiston. In Louisiana and Texas, sugarcane varieties are obtained from a cooperative arrangement between the USDA-ARS Sugarcane Field Station at Canal Point, Florida, the USDA-ARS Sugarcane Field Laboratory at Houma, Louisiana, and the Texas Agricultural Experiment Station at Texas A & M University, Weslaco, Texas. True sugarcane seed is produced at the USDA-ARS Sugar Cane Field Station at Canal Point, Florida, with actual field testing conducted by researchers in Louisiana and Texas. Louisiana also obtains varieties from the sugarcane breeding program located at the Agricultural Experiment Station of the Louisiana State University Agricultural Center at Baton Rouge. Sugarcane varieties grown in Hawaii are developed by the Hawaiian Sugarcane Planters' Association.

These breeding programs are very expensive because they require large numbers of highly trained personnel, elaborate breeding facilities, and large amounts of land for field testing. Generally, researchers with advanced degrees in genetics, pathology, and agronomy head the breeding programs, and they are assisted by personnel with B.S. degrees in agronomy or related disciplines. In addition, considerable field labor is needed for the breeding, planting, sampling, and harvesting operations.

Each plant that develops from a true seed is genetically different from all others and is a potential new variety. The various varietal development programs evaluate from 20,000 to over 500,000 of

Figure 2-3. Sugarcane elevator with cane being loaded from truck to railcars. (Source: Henry J. Andreis, U.S. Sugar Corp.)

these plants each year with the goal of finding one or two that will be suitable for commercial production.

Among varietal characteristics breeders screen are early ripening, high-sucrose content, high sugarcane tonnage, strong restubbling ability, disease resistance, insect resistance, erectness, low-fiber content, and cold tolerance. In general, it takes approximately ten years for a breeding program to release a new variety acceptable to the industry.

Sugarcane Diseases. Diseases cause some of the most serious problems associated with growing sugarcane. Most diseases can only be controlled by developing varieties resistant to the diseases present in each sugarcane region. It is advantageous to grow several varieties, some of which may be slightly lower in yield or sugar content, to provide sufficient genetic diversity in the crop to minimize major losses should a new pathogen be introduced into an area. Large losses have occurred in countries overly dependent on one variety that later proved to be susceptible to a new disease. Serious diseases that have been controlled by developing resistant sugarcane varieties are mosaic virus, rust, and smut. While these diseases have not been eliminated from the U.S. sugarcane-producing areas, their detrimental effects on sugar production have been greatly reduced. Ratoon stunting disease, one of the most widespread bacterial diseases of sugarcane in the world, can be controlled by treating the seedcane with hot water prior to planting and using sanitation practices to prevent the reintroduction of the disease. A few diseases, such as pineapple disease, which is caused by a fungal soil pathogen, can be controlled by applying a fungicide prior to planting.

Sugarcane Insect Pests. One of the most serious insect pests of sugarcane is the sugarcane borer, *Diatraea sacharalis*. This sugarcane borer is a problem in all four sugarcane producing states. Borers can attack young plants near the growing points, causing their tops to die, but the most common damage is done when they bore into stalks and feed on the soft sugar-containing tissue, thereby reducing cane tonnage, juice quality, and sugar yields. Other important insect pests found in one or more of the U.S. sugarcane areas include grubs, aphids, mites, army worms, the southwestern corn borer, the lesser cornstalk borer, and the rice stem borer. Depending on their population levels, each of these pests can cause losses in sugar production. Some aphids and planthoppers transmit sugarcane diseases.

While insecticides are an important insect-control tool, control by cultural practices and natural control have been practiced by the industry for many years. Cultural practices such as repeated discing or flooding of fields help control soil insects. Burning reduces the trash that harbors insects. Insect-susceptible varieties are generally avoided, and research is underway to develop new insect-resistant varieties, notably against the sugarcane borer. Biological control tactics for sugarcane insects, which involve the utilization of natural or released parasites, predators, and pathogens that feed on the pest insects, have been exploited in each of the four U.S. sugarcane areas. Integrated pest management systems involving cultural, varietal, biological, and insecticidal control tactics in conjunction with specific control-threshold levels and scouting programs are being developed and implemented.

Raw Sugar Processing

Raw sugar mills vary in size from those capable of processing 2,500 tons of cane per day to those that can process over 24,000 tons of cane per day. A simplified flow diagram of a raw sugar mill is illustrated in Figure 2-4.

After the cane is delivered to the mill yards, excessive soil and rocks are removed, if necessary, prior to processing. Cane entering the mill is prepared by chopping, shredding, or crushing, and after one or a combination of these procedures is completed, the remaining juice is extracted by passing the prepared cane through a series of mills containing three to five rollers. Figure 2-5 shows shredded sugarcane entering the first in a series of three-roll mills. As the cane is crushed, hot water or a combination of hot water and recovered, impure juice is sprayed onto the crushed cane as it leaves each mill to aid in the extraction of sugar. In many factories bagasse, or plant residue, is sent to the boilers to be burned as fuel after it exits the last mill.

Juice from the mills is quite turbid and contains considerable impurities. To remove both the insoluble and the soluble impurities from the juice, the juice is limed, heated, and sent to clarifiers.

In the clarifiers, the addition of lime and heat causes some of the suspended impurities to coagulate and entrap both soluble and insoluble impurities, producing a precipitate that settles to the bottom of the clarifiers. The clarified juice is sent to evaporators, while the juice-containing precipitate is sent to rotary-drum vacuum filters to extract the juice, leaving the filter cake for disposal. The filtered juice is returned to the raw-juice line for additional treatment, while the filter cake is generally applied to nearby sugarcane fields for its soil building and nutritive value or stored for later use.

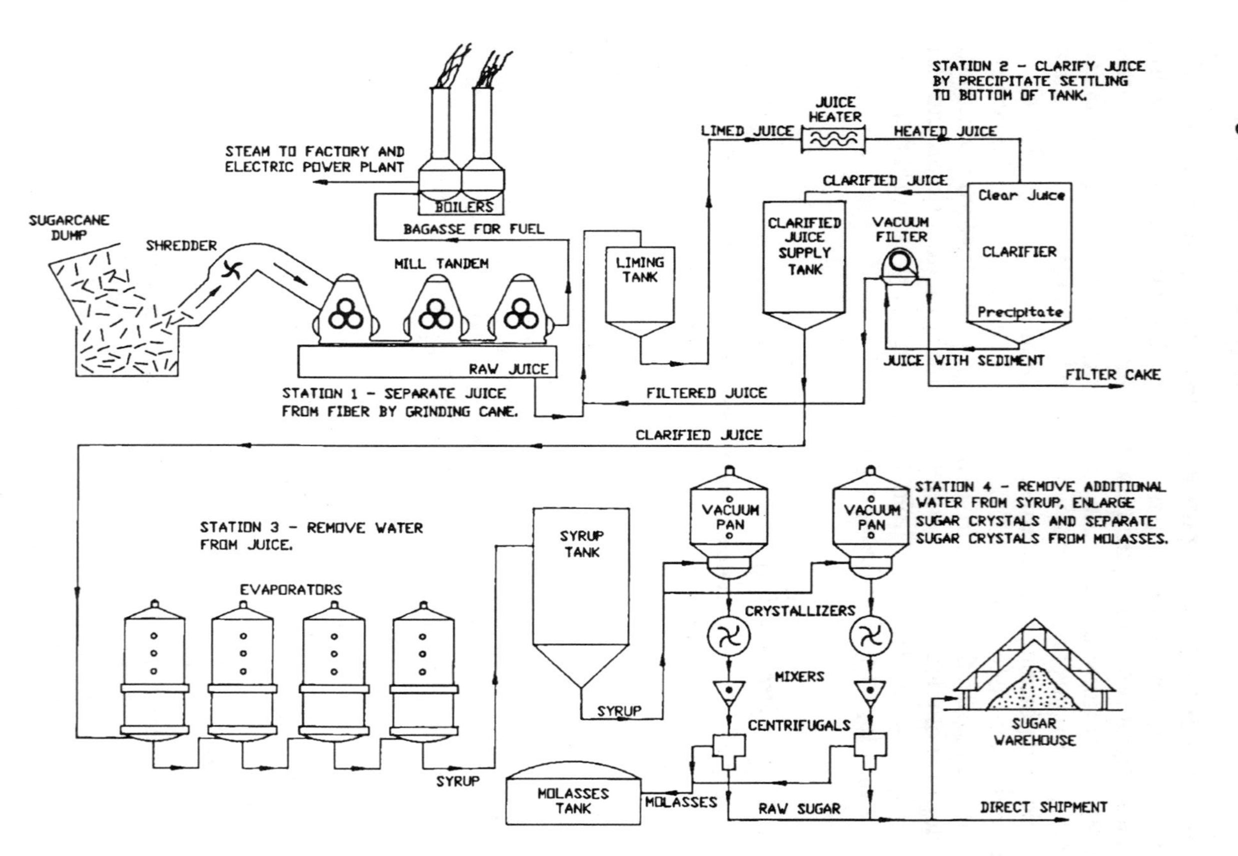

Figure 2-4. Simplified flow diagram of a raw sugar mill. (Prepared by Henry J. Andreis and Kevin Hook, U.S. Sugar Corp.)

Clarified juice commonly contains about 85 percent water and is concentrated by heating in vacuum evaporators to remove approximately two-thirds of the water. The water is evaporated in a continuous series of three or four vacuum-boiling cells. The syrup that leaves the last vacuum boiler contains about 35 percent water and 65 percent solids.

This syrup is further concentrated by additional evaporation in the crystallizer vacuum pans until saturated with sugar. Seed grains (small grains of sugar) are added to serve as nuclei for the formation of sugar crystals, a process known as crystallization. Also, additional syrup is added and evaporated, so that the original crystals that were formed are allowed to grow in size. When sucrose concentration reaches the desired level, the dense mixture of syrup and sugar crystals, called massecuite, is discharged into large containers known as crystallizers. Crystallization continues in the crystallizers as the massecuite is slowly stirred and cooled. At the appropriate time, the mixture of nearly exhausted syrup and sugar crystals is sent to mixers where the crystals are prevented from settling while additional cooling takes place. Finally, massecuite from the mixers is allowed to flow into centrifugals where the very

Figure 2-5. Shredded sugarcane entering the first in a series of three-roll mills. (Source: Sugar Cane Growers Cooperative of Florida)

thick syrup, known as molasses, is separated from the raw sugar by centrifugal force. The raw sugar is retained in the centrifuge baskets on a fine screen while the final molasses passes through the screen perforations. Figure 2-6 shows raw sugar leaving the centrifugals and being delivered by conveyor belt to trucks for transport to a warehouse. The final molasses (blackstrap molasses), containing sucrose, reducing sugars, organic nonsugars, ash, and water is sent to large storage tanks. The blackstrap molasses is used as a cattle feed and in the production of industrial alcohol, yeast, organic chemicals, and rum.

Raw sugar generally contains 96 to 99 percent sucrose, with small amounts of water, invert sugar, and minerals present. The invert sugar, water, and minerals are found primarily in a very thin film of molasses that covers the crystals. The raw sugar is sent directly to refineries to be processed into refined sugar or is stored in large warehouses to be sent to refineries at a later date (Figure 2-7).

More information on all aspects of sugarcane culture and raw sugar production can be obtained from the standard reference works (4–10) cited in the reference list.

References

1. Rosenfeld, A. H., Jr. *Sugar Cane Around the World*, pp. 1–7. University of Chicago. 1956.
2. Sitterson, C. J. *Sugar Country*, pp. 1–12. Kentucky Press, Lexington, KY. 1953.
3. USDA Economic Service. *Sugar and Sweetener Situation and Outlook*, p. 25. U.S. Government Printing Office, Rockville, MD. 1989.
4. Humbert, R. P. *The Growing of Sugar Cane*. Elsevier, New York. 1968.
5. Barnes, A. C. *The Sugar Cane*. Interscience Publishers, New York. 1964.
6. Van Dillewijn, C. *Botany of Sugarcane*. Chronica Botanica, Waltham, MA. 1952.
7. Martin, J. P., Abbott, E. V., and Hughes, C. G. *Sugar-Cane Diseases of the World*. Vol. I. Elsevier, New York. 1961.
8. Hughes, C. G., Abbott, E. V., and Wismer, C. A. *Sugar-Cane Diseases of the World*. Vol II. Elsevier, New York. 1964.
9. Meade, G. P. *Cane Sugar Handbook*. John Wiley and Sons, New York. 1977.
10. Williams, J. R., Metcalfe, J. R., Mungomery, R. W., and Mathes, R. *Pests of Sugar Cane*. Elsevier, New York. 1969.
11. Hall, D.G. "Insects and Mites Associated with Sugarcane in Florida." *Fla. Entomol.* 71:138–150. 1988.
12. Long, W. H., and Hensley, S. D. "Insect Pests of Sugarcane." *Ann. Rev. Entomol.* 17:149–176. 1972.

Figure 2-6. Raw sugar leaving the centrifugals on a conveyor belt. (Source: Sugar Cane Growers Cooperative of Florida)

Figure 2-7. Raw sugar being piled in a sugar warehouse. (Source: Sugar Cane Growers Cooperative of Florida)

Cane Sugar Refining

A. B. Rizzuto *

The origins of modern cane sugar refining can be traced to the seaports of Great Britain and Western Europe during the sixteenth century. The process of producing refined sugar at that time was primitive and wasteful compared to current methods. Raw sugar loaves were dissolved and heated in open kettles with lime water. Egg albumin was then added to the heated juice and the precipitated coagulum of impurities removed by skimming and filtering through cloth. The liquid was then evaporated to a thick syrup, during which time crystals formed. This mixture was poured into conical molds from which the adhering syrup or molasses drained into a pot through the opening at the mold point.

The sugar in the mold was then covered with a pasty mass of clay and water, which percolated downward to remove most of the residual molasses. After draining, the loaves of sugar were removed from the molds and dried in ovens, the product thus obtained being a single-refined loaf.

The single-refined product was frequently stained with yellow patches due to traces of unremoved molasses. To remove this defect, the loaves were remelted and run back through the process. This second crystallization and baking resulted in a double-refined loaf. Frequently, this product was pulverized and sold as powdered sugar. The drippings of molasses and syrup from the molds were then boiled and crystallized to produce "shop-sugars," which were graded according to differences in color. The process for refining sugar remained essentially the same through the end of the eighteenth century.

By the nineteenth century the Watt steam engine was providing a dramatically improved source of power and heat for sugar refining. In addition, the vacuum pan was invented in 1813 by the English sugar refiner Howard. Its use prevented the destruction of sugar from overheating in open evaporating kettles. A French pharmacist, Figuier of Montpelier, was making boot polish from charcoal and honey in 1812, when he noted that charcoal decolorized honey. Derosne, in 1827, is credited with the development of the filtration of sugar solutions through bone char columns. Another advance

*A. B. Rizzuto is Director of Research at Amstar Sugar Corporation, Brooklyn, NY.

occurred in 1836, when Degrand designed the double-effect evaporator. Norbert Rillieux, a native of New Orleans, Louisiana, invented the multiple-effect evaporator in 1832 and applied for his first two patents in 1843.

The labor needed to handle sugar molds, the large amount of space the molds occupied, and the long time needed for drainage of molasses from the molds required large capital expenditures. These problems were simplified in 1837, when Penzoldt invented the centrifugal machine. Although intended for use in the wool industry, it was used for purging sugar in 1843 by Hardman. In 1850, Sir Henry Bessemer introduced a steel-framed centrifugal that was suspended from a ball-and-socket joint. The suspended centrifugal machine enabled refiners to produce high-quality granulated sugar easily and cheaply.

The first successful filter press for sugar is credited to Needham in 1853. Prior to that time, thick layers of scum settled on the surface of clarified juice. Skimming and settling were required to remove the insoluble impurities, which increased time and labor, and was accompanied by a substantial loss of sugar. The filter press made it possible to remove the impurities of clarified juice with neatness and dispatch.

These inventions—the steam engine, steam-heated vacuum pans, the bone char process, multiple-effect evaporators, press filtration, and the centrifugal machine—greatly improved the process and lessened the cost of sugar refining. By the second half of the nineteenth century, the number of refineries and consequent availability of refined sugar began to increase.

Cane Sugar Refining Today

A review of twentieth century cane sugar refining shows little basic change, but the efficiency of each process has been optimized. Today, the process still employs five basic unit operations:

1. Centrifugation
2. Filtration
3. Decolorization
4. Evaporation
5. Crystallization

(See definitions in Chapter 18, Terminology.)

The refining process is essentially a laundry-type operation. Its basic objective is to launder sucrose using the above-mentioned five unit operations. Using these unit operations in concert, like the working of the fingers on a hand, the process provides pure white cane sugar (99.96 percent sucrose) and a residual blackstrap (molasses). Molasses is a collection of concentrated nonsugar solids plus a limited amount of sucrose retained during processing.

Raw sugar enters the process containing nonsugar material on its crystal surface. The first step in the process separates the surface molasses film from the relatively pure raw sugar crystal. This is accomplished by mingling hot, saturated syrup with raw sugar followed by centrifugation. At the centrifugal machine, the liquid affination syrup and the washed raw sugar are separated. Affination syrup is reused in the mingling process, with the excess sent to a recovery system for final sucrose extraction and production of molasses.

Concurrently, washed raw sugar is dissolved in high-purity sweetwater to produce a high-density washed sugar liquor. The liquor is then treated with lime and carbon dioxide, lime and phosphates, or other chemicals. This treated liquor is then clarified by use of flotation clarifiers or is passed through a filtration apparatus that removes the colloidal particles and produces a sparkling golden sugar liquor.

Decolorization is generally accomplished by passing the press-filtered, washed sugar liquor through a decolorizing column containing a charcoal-like material that selectively removes soluble color bodies while allowing sucrose to pass through unchanged. The decolorized, charcoal-filtered liquor is then passed through a multiple-effect evaporator that removes water, presenting the vacuum pan with a highly concentrated sugar solution. In white sugar boiling, the vacuum pan continues the evaporation process until a seeding procedure using tiny sucrose crystals shocks the supersaturated sugar solution into producing an initial crop of sugar crystals. By control of the rate of evaporation, these crystals continue to grow until they reach a predetermined size. At this time, the crystals, in a sea of mother liquor called massecuite, are transferred from the vacuum pan to a mixer that feeds the mixture to centrifugal machines where, once again, the syrup is removed from the pure sugar crystals. Steam is used to wash the crystals in the centrifugal before they are sent to driers. The syrup is moved through successive boilings to produce consecutive crops of sugar crystals. The damp sugar crystals are dried by tumbling through heated air in a granulator. After the crystals are dried, they are screened, classified, and placed in various holding bins. Packaged, bulk, and liquid sugar are produced for sale from these source bins.

The process diagram in Figure 2-8 illustrates the separation technology employed in a cane sugar refinery.

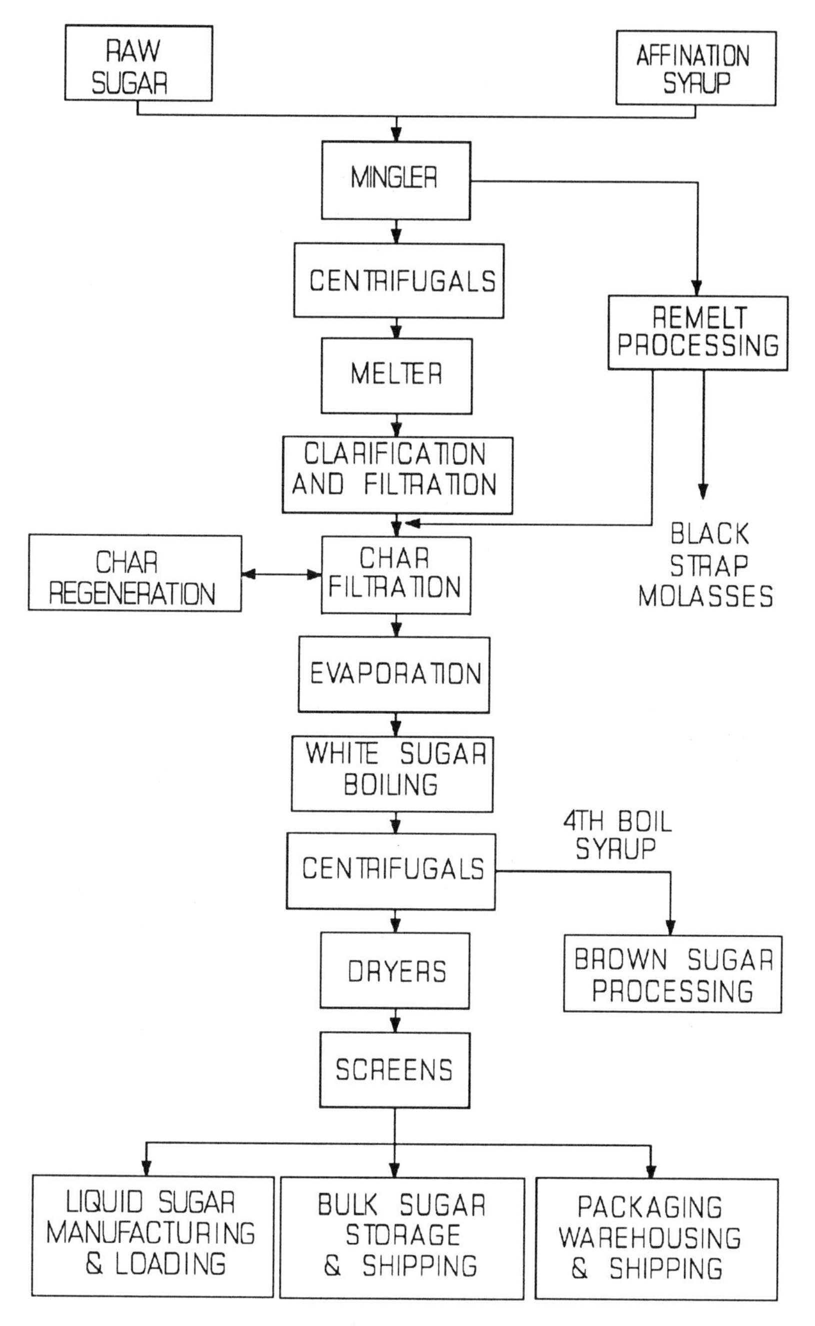

Figure 2-8. Cane sugar refining process flowchart. (Source: Amstar Sugar Corp.)

Beet Sugar Production

S. E. Bichsel *

The United States beet sugar industry developed from the European beet sugar industry, which began in France and Germany in the early 1800s. Today, sugar beets are produced in twelve states located in the western and central parts of the United States (Figure 2-9). Since the inception of the U. S. beet industry in 1900, dramatic increases in yield per acre and harvested acres of sugar beets have occurred, from a minimal amount in 1900 to approximately 1.3 million acres harvested in 1989 at an average yield of 19.0 tons per acre (Figure 2-10). The dramatic increase in land dedicated to sugar beet production and the increase in productivity per acre have resulted primarily from sugar price stability, advances in cultural practices, improved varieties of beets, and larger, more cost-effective sugar beet receiving stations, storage, and processing plants. The availability of disease-resistant, monogerm hybrid varieties of sugar beets has helped overcome common sugar beet diseases and has reduced labor costs associated with thinning the original multigerm varieties. The combination of stable beet sugar costs, technological advances in sugar beet production, and more efficient beet sugar processing has promoted the growth of the domestic beet sugar industry.

Refined sugar production per acre has increased significantly from the mid-1960s through the mid-1980s (Figure 2-11). During this period of the domestic beet sugar industry's history, continuous gains were made in agronomic practices and hybrid sugar beet variety development. Improvements in refined sugar processing enabled the beet sugar industry to maintain refined sugar production in spite of a decline in harvested acres. Approximately 60 million hundredweight of domestic beet sugar was produced per year from the mid-1960s through the mid-1980s (Figure 2-12).

Sugar Beet Cultivation

Using a precision planter, sugar beet monogerm seed is planted from March to May as a row crop (generally 22-inch rows). Seed spacing varies generally from 3 to 5 inches to obtain optimum plant popu-

---

*S. E. Bichsel is Senior Vice President Research and Development, Holly Sugar Corporation, a subsidiary of Imperial Holly Corporation, Colorado Springs, CO.

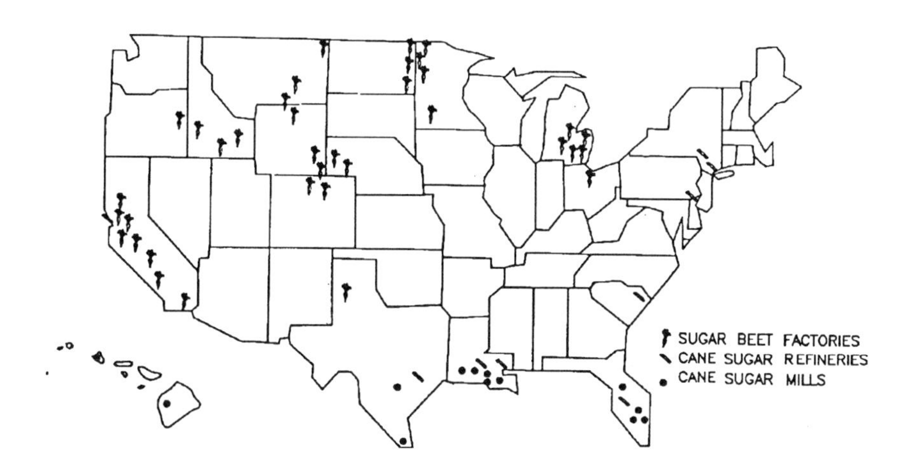

Figure 2-9. North American beet and cane sugar manufacturing facilities. (Source: S. E. Bichsel)

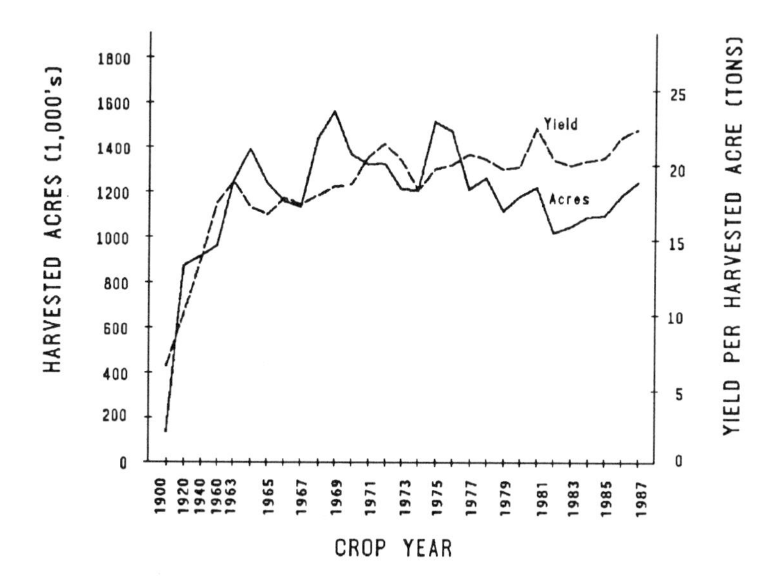

Figure 2-10. Harvested sugar beet acres and yield per acre by year. (Source: U.S. Beet Sugar Association, Washington, DC)

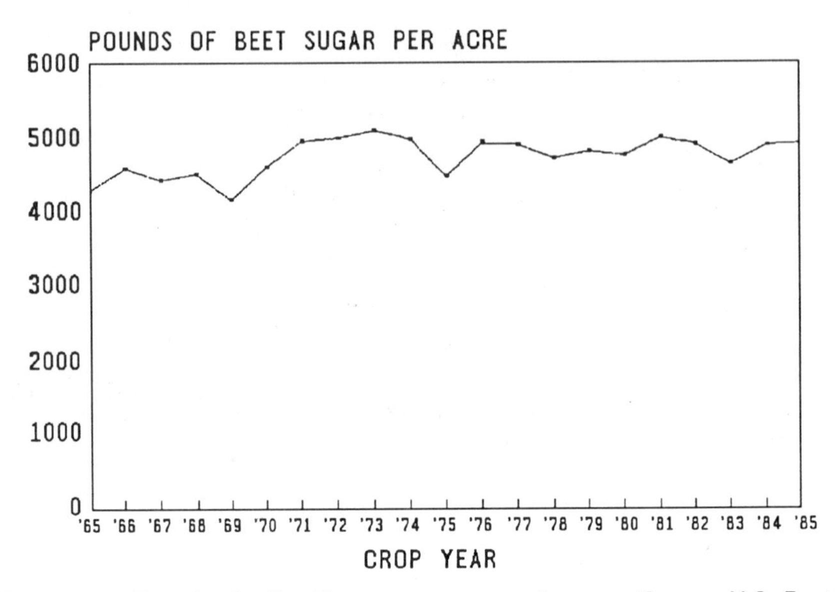

Figure 2-11. Pounds of refined beet sugar per acre by year. (Source: U.S. Beet Sugar Association, Washington, DC)

lation per acre. Seeds are planted from ¾- to 1¼-inch deep in soil that has been prepared to ensure a firm yet porous seed bed for optimum germination. Registered pre- or post-emergence herbicides are utilized to minimize hand labor associated with weed control. Sugar beets are grown on irrigated land primarily in lighter western soil and on nonirrigated land in heavier, higher-organic soil in Minnesota, North Dakota, Ohio, and Michigan. Beet seedlings are generally thinned at the six-leaf stage to optimum spacing for maximum yield and root quality. Approved chemical sprays may be necessary at critical times during the sugar beet growing season to minimize the possibility of airborne fungus diseases that can damage the sugar beet leaf and significantly reduce yield and quality of the harvested root (1). Prior to harvest, sugar beets are defoliated using multirow defoliaters to remove all top growth from the crown of the beet. Sugar beet top petiole and leaf material are incorporated into the soil as organic matter or fed to livestock. During or after defoliation, sugar beets are generally scalped to eliminate low-quality top crown tissue before harvest. Sugar beets are generally harvested from September through the first part of November with multirow lifting-wheel harvesters. In central and northern California, a designated percentage of the sugar beet crop is allowed to

overwinter for harvest in the spring. Winter temperatures do not generally fall below freezing in these areas. Beet storage in the ground extends total time of factory operation per year and contributes to economic return on factory capital investment.

Sugar Beet Long-Term Storage

In California, harvested sugar beets are delivered directly to the processing factory. Normal storage prior to slicing does not exceed forty-eight hours. An exception is short-term stockpiling in the case of fall or spring rains to maintain uninterrupted factory operation. In the inland western, north central, and midwest sugar beet growing areas of the United States, harvesting and processing of sugar beets proceeds in unison until the first part of October. Beyond that time, beets will be harvested at a rate in excess of factory processing capacity. Harvest is generally completed by mid-November to avoid irreversible freeze damage of sugar beets remaining in the ground after mid-November. The excess beets are stored in piles generally 180 to 200 feet wide and 20 to 24 feet high. These piles are located at the processing plant or in piling grounds remote from

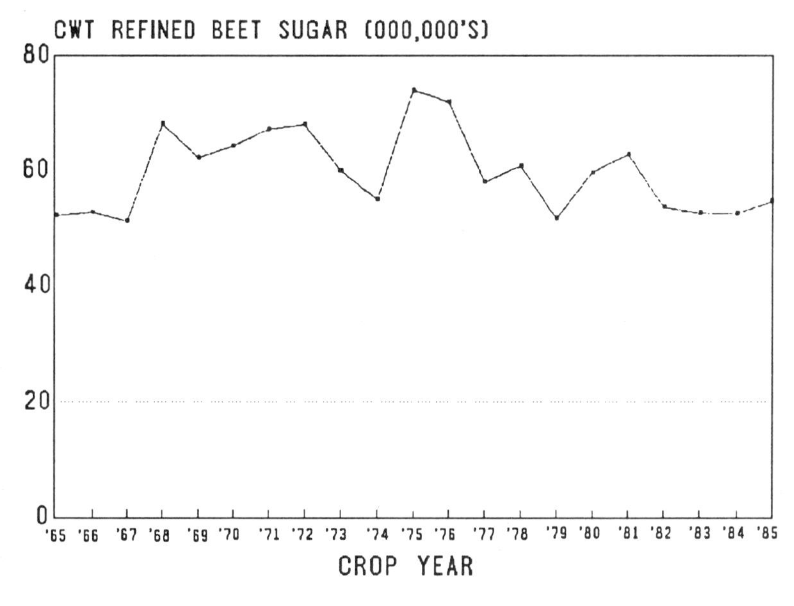

Figure 2-12. Refined beet sugar production (million hundredweight) by crop year. (Source: U.S. Beet Sugar Association, Washington, DC)

the factory location. Beets are generally piled in long-term storage from 80 to 160 days, depending upon average winter ambient temperatures. Sugar loss in stored sugar beets, due to live root respiration and external and internal microorganism-induced tissue rot, is specifically a function of storage time, root temperature, and sugar beet variety. Under ordinary storage conditions, a sugar loss of ½ pound per ton per day may be assumed. Every effort to induce natural circulation of cool air through the beet piles is promoted. In some areas, ventilation fans and covered-air distribution channels are used to force cool outside air through the beet pile. The purpose is to reduce root temperatures to a few degrees above freezing, at which point sugar loss due to respiration is reduced to a minimum. A 10°F increase in root temperature more than doubles the root respiration and subsequent sugar loss. In the north central area, located in North Dakota and Minnesota, arctic-type winter temperatures that allow deep freezing of sugar beet roots result in a total cessation of sugar loss due to respiration and microorganism attack.

Sugar Beet Processing

Sugar beet processing is illustrated in Figure 2-13. The complete sugar beet processing operation involves a number of unit operations, including the following:

1. Diffusion (Dried beet pulp—animal feed co-product)
2. Carbonation (Purification)
3. Evaporation (Concentration)
4. Crystallization (Salable white sugar; also molasses co-product used as animal feed and fermentation feed stock)

Sugar beets from the storage piles are moved into the factory, washed, and sliced into V-shaped cossette sections measuring approximately 2⁄16-inch on a side and 2 to 3 inches in length. This shape of cossette ensures maximum surface area for sugar diffusion into the hot aqueous extraction solvent. The cossettes are introduced into a countercurrent diffuser that may be a horizontal drum, sloped vat, or vertical-cylindrical tower (2). All countercurrent continuous diffusion units involve a mechanism to transport sugar beet cossettes in a direction against the flow of hot water extractant. Sugar beet cell walls are heat-denatured to enhance diffusion of sugar from the plant cells to the lower concentration of sugar in the

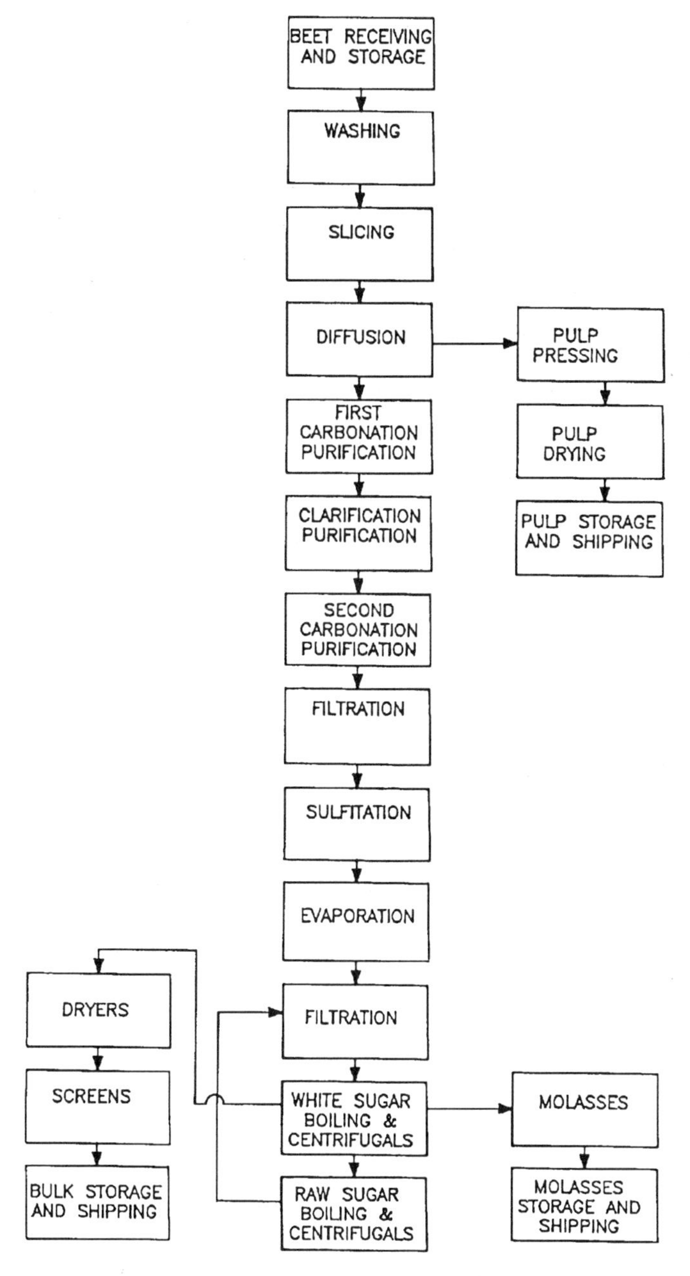

Figure 2-13. Beet sugar process flowchart. (Source: Holly Sugar Corporation, a subsidiary of Imperial Holly Corporation)

aqueous extractant. Approximately 98 percent of the sugar is extracted in the countercurrent diffusion operation. Wet beet pulp, discharged from the diffuser at approximately 92 percent moisture, is pressed mechanically to reduce beet pulp moisture to approximately 76 percent, and either air-dried under ambient temperatures in California or dried in direct gas, oil, or coal-fired rotary drum dryers to a moisture content of approximately 10 percent prior to sale as an animal feed. Juice from the diffuser, containing approximately 12 percent sugar by weight and 2 percent soluble impurities in addition to soluble and semisoluble colloidal proteins, pectins, and saponins extracted from the cossettes, is heated to 85°C prior to lime-carbon dioxide purification.

Clarification of diffusion juice involves coagulation and precipitation of the semisoluble colloidal materials as insoluble salts, as well as precipitation of the potassium and sodium salts of oxalic, malic, and citric acids, and phosphates and sulfates as insoluble calcium salts (3). Continuous lime and carbon dioxide purification involves concurrent addition of milk of lime and carbon dioxide to diffusion juice in a carbonation tank. The retention time in the carbonation tank is approximately twelve to fifteen minutes under ideal temperature conditions of 85° to 88°C.

The treated juice is then sent to a clarifier for separation of the precipitated impurities. Thickened underflow mud from the clarifier is generally filtered and washed on rotary vacuum drum filters to minimize sugar loss.

Lime and carbon dioxide are produced from limerock at each factory in a coke or natural-gas-fired vertical-shaft Belgian-type lime kiln, or in a gas-fired horizontal rotary drum-type kiln where waste lime is reburned. Carbon dioxide is reclaimed from the kiln for direct use in the lime-carbon dioxide first carbonation purification tank. Approximately 2 percent lime is used on sugar beets for purification and clarification.

Clarified overflow juice is then subjected to a second treatment with carbon dioxide to reduce residual lime salts concentration. The calcium carbonate precipitate is filtered from the juice by use of pressure filters. Sulfur dioxide is added to the filtered juice to minimize color formation during subsequent processing steps. The thin juice, after all colloidal impurities and approximately 35 percent of the soluble impurities have been removed, is concentrated from about 13 percent dissolved solids to approximately 60 to 65 percent dissolved solids in energy-efficient quintuple-effect evaporators.

Steam at all beet sugar factories is produced in relatively low-pressure boilers, 250 to 400 pounds per square inch. This steam is passed through a turbine to produce electricity for factory use. The exhaust steam from the turbine is used for heating, evaporation, and crystallization of sugar in the latter stages of the process.

Thick juice from the evaporators is concentrated and sugar is crystallized in a three-stage process. The first vacuum-pan crystallization produces white sugar for direct sales. The second and third vacuum-pan crystallizers produce lower-purity raw sugars, which are redissolved after separation of the massecuite mother liquor. Continuous conic-basket centrifugals, developed specifically for the sugar processing industry, are used for further purification of this lower-purity raw sugar. Salable white sugar is produced by crystallization in a vacuum pan from a thick feed juice of approximately 92 percent purity (92 percent of the dissolved solids are sugar, 8 percent are nonsugars). Crystallized sugar and massecuite mother liquor are separated in basket-type batch centrifugals prior to washing of the recovered crystals with water.

The white sugar, discharged wet from the batch centrifuges, is dried in rotary hot-air dryers (granulators) to a moisture content of approximately .02 percent. The white refined sugar produced is 99.96 percent pure as a result of successive purification by diffusion, lime-carbon dioxide purification, and triple crystallization in the batch vacuum-pan crystallizers.

The mother liquor recovered from the last of the three vacuum pan crystallizers is exhausted of further raw sugar crystals under standard atmospheric conditions. The massecuite liquor reclaimed from this process is molasses co-product, and is approximately 60 percent pure. The ratio of nonsugar impurities to sugar prevents further crystallization of sugar using conventional processing technology.

Summary

As indicated in Figure 2-14, the objective of sugar beet processing is to separate pure sugar from insoluble pulp, soluble nonsugars, and water at a minimum cost. A typical harvested sugar beet contains 75.9 percent water, 2.6 percent soluble nonsugars, 16.0 percent sugar, and 5.5 percent pulp, which is the insoluble cellulose, hemicellulose, and pectin substances remaining after sugar extraction from cossettes. Figure 2-15 indicates the material recovery from one ton of 16 percent sugar-by-weight beets. In quantitative terms of white sugar, 266 pounds of sugar per ton of beets processed

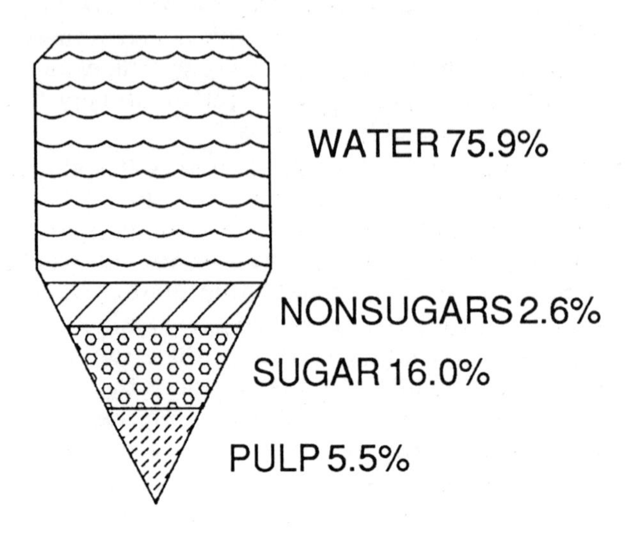

Figure 2-14. Harvested sugar beet composition percent on total weight. (Source: S. E. Bichsel)

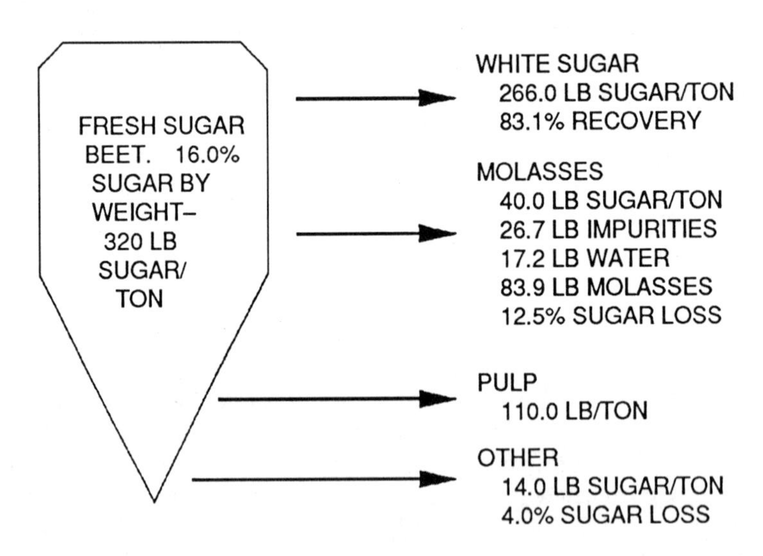

Figure 2-15. Sugar recovered, sugar lost, and by-products produced from one ton of fresh sugar beets.(Source: S. E. Bichsel)

is produced for sale, or 83.1 percent of the total sugar is extracted. As shown in Figure 2-15, there are 26.7 pounds of impurities, 17.2 pounds of water, and 40.0 pounds of sugar in the molasses by-product for each ton of beets (16.0 percent sugar) processed. The 40.0 pounds of sugar represent a 12.5 percent loss of the original weight of sugar in the fresh sugar beets. The 83.9 pounds of recovered molasses is 60 percent pure (percent of the total dry weight that is sugar). In addition to the 83.9 pounds of molasses by-products produced, 110.0 pounds of beet pulp are produced per ton of fresh sugar beets processed. Other sugar losses include sugar remaining in pulp after diffusion, in the lime cake after washing, and in the rotary vacuum-drum filters, and sugar lost because of bacterial action, sugar-containing liquid spills, and sugar inverted and caramelized following exposure to hot heating surfaces during processing.

References

1. McGinnis, R. A. *Beet-Sugar Technology.* Beet Sugar Development Foundation, Ft. Collins, CO. 1982.
2. Vukov, K. *Physics and Chemistry of Sugar Beets in Sugar Manufacture.* Elsevier, New York. 1977.
3. Silin, P. M. *Technology of Beet-Sugar Production and Refining*, p. 47. Tr. L. Markin. National Science Foundation, Washington, D. C., 1964.

3

Sugar Products

J. F. Dowling*

Sugar products produced and sold in the United States can be divided into four basic categories: (1) Granulated, (2) Liquid, (3) Brown Sugars, and (4) Specialty Products. These sugar products will be reviewed, incorporating typical properties of both beet and cane sugar and products found in various sections of the U.S. Sugars requiring special properties, such as product-specific purity and unique grain sizes, are available from various refiners. Information about specific sugar products and special requirements for a particular application should be obtained from a sugar supplier.

Granulated Sugar Products

Granulated sugar, one of the oldest and purest flavors used in the food industry, is essentially pure crystalline sucrose. Standard granulated sugar products sold by both beet and cane refiners will contain at least 99.8 percent sucrose. These products are processed to a variety of crystal-size distributions to meet the requirements of the various applications. Table 3-1 offers a summary of the typical analysis for the basic granulated sugars produced today.

Granulated sugars are normally defined by the percent of the crystals retained on a particular standard U. S. mesh screen. This percent can vary with crystallization and refinery screening methods.

Coarse granulated sugar is normally crystallized from the highest-purity sugar liquor. Thus, it is the highest-purity crystalline product with the lowest color and best color stability. It is used in fondants, liquors, and other formulations where colorless white products are required.

*J. F. Dowling is Technical Manager, Refined Sugars, Inc., Yonkers, NY.

Sanding granulated sugar, which normally ranges between U.S. 20- and U.S. 40-mesh screens, is used mainly by the baking and confectionery industries as a sprinkle on baked goods or sanding for starch gums and similar items. Its grain size is such that the crystals reflect light and give the product a sparkling appearance.

Extra Fine (X-Fine) or Fine sugar has grain sizes that fall between U.S. 20- and U.S. 100-mesh screens and is the largest-volume industrial granulated sugar sold today. Its particular size makes it ideal for bulk handling and less susceptible to caking. It can be used in dry mixes where a larger grain, or lower-crystal-surface area, is desired.

Fruit granulated sugar is slightly finer than X-Fine, with a grain-size distribution between U.S. 40 and U.S. 100 screens. This product, because of its smaller grain and larger surface area, is used mainly in dry mixes such as gelatin desserts, pudding mixes, and drink mixes. Its narrower grain size distribution lends itself well to less stratification, or separation of ingredients, in formulated dry mixes.

In dry mixes, sugar grain size will contribute to the color pickup of the mix. Basically, the larger the grain size the less the surface area, and therefore the greater the color-additive pickup per crystal. As the crystal size becomes smaller, there will be a larger surface area per constant weight and thus less color pickup compared with the same weight of larger crystals.

If the crystal distribution in a particular product is large, there could be settling in the mix with the fines separating to the bottom. In mixing ingredients it is important that the crystal sizes be of similar nature to prevent separation.

Bakers Special, as the name implies, was developed mainly for the baking industry. With a finer-grain-size distribution between U.S. 50- and U.S. 140-mesh screens, it lends itself to sugaring of cookies and doughnuts. The fine-grain size is also helpful in producing a good cell structure in cakes. Since its crystal size is the smallest of the standard granulated products, Bakers Special sugar dissolves the fastest of all granulated sugars, and is often used to sweeten bar drinks.

Normal granulated sugars are available in 50- and 100-pound bags. Granulated sugar is also sold in bulk via rail or truck.

The color of granulated, crystalline sugars is essentially white to the eye, but when dissolved in water may appear pure white to light yellow. Normally, sugar colors are measured by procedures approved by the International Commission for Uniform Methods of

Sugar Analysis (ICUMSA). ICUMSA method 4 specifies membrane (.45 micron) filtration of a 50 percent solution and measurement at 420 nm. The absorbency is converted to International Color Units (ICU), where the higher the ICU number the darker the color. Most granulated sugars fall below the 35 ICU absorbency maximum. This limit was established by the carbonated beverage industry to meet the color requirements of water-white beverages. The percent ash, or an estimate of inorganic content of the sugar, is normally measured by conductivity because of the extremely low value. Most sugars will average about 0.02 percent (Table 3-1).

The percent moisture for granulated sugars is also very low (below 0.04 percent) (Table 3-1) as most products are "conditioned" after drying. Most sugar processors will condition or age sugar to allow its moisture content to come to equilibrium with the surrounding atmosphere. There are basically three types of moisture associated with granulated sugar. The first type is termed "included," and refers to syrup trapped within the growing sugar crystal during the crystallization stage. This moisture (syrup) can only be liberated when the crystals are ground (broken). The second type of moisture is referred to as "bound," and is a film of saturated syrup surrounding the outside of the crystal. When sugar is dried rapidly above room temperature, a supersaturated solution is formed and is deposited on the surface of the crystals. As the sugar cools and ages, crystallization takes place in this syrup phase and liberates water. This excess water is the third type of moisture, and is referred to as "free moisture" since it can escape into the surrounding air. The process of going from bound to free moisture normally requires twenty-four to forty-eight hours. Thus, sugar processors will condition (store in moving silos) sugar prior to packaging to allow for moisture stabilization.

If free-flowing, dry sugar is exposed to high humidity, it will tend to pick up moisture with syrup forming on the crystal surfaces. When the humidity declines, crystallization reoccurs and caking (hardening of the sugar) will follow.

This is normally what happens when one puts dry sugar in a sugar bowl not totally dry and finds that the sugar has become hard the next day. Control of moisture in refined sugar permits the shipping and handling of granulated sugars in bulk. Granulated sugars have low water activity and, therefore, are microbiologically stable. Typical granulated sugar products comply with the carbonated beverage industry specifications (3) of not more than 200 mesophilic bacteria, 10 yeast, or 10 mold per 10 grams of sugar. Most granulated sugars also comply with standards established by the National Food Processors Association (2) and the pharmaceutical industry (U.S. Pharmacopeia, National Formulary) (7).

Powdered confectioners sugar is produced by grinding or milling a mixture of granulated sugar and corn starch. The starch (normally 3 percent) is added to absorb the freed moisture and prevent the sugar from caking. These sugars have the smallest particle size of any crystalline sucrose product and are used for dusting in the dry form or to form wet fondants when mixed with water. The microcrystals produce a smooth fondant. Powdered sugars are sold in 50- and 100-pound bags. Although various grades of powdered sugars are produced, the two most common are 6X and 10X powdered sugar (Table 3-1). The higher the number prior to X, the finer the powdered sugar. In the production of fondants, the finer the crystal size, the less gritty the taste and the smoother the fondant. The low color content of these powdered sugars allows for water-white fondant mixes.

Liquid Sugar Products

Liquid sugar products were developed before the process of granulated sugar conditioning and the installation of bulk handling systems. They were easily transported, handled, and measured within industrial plants. There are four basic types of refined liquid sugar available today (Table 3-2).

Liquid sucrose, which is essentially a liquid granulated sugar, is produced at minimum concentrations of 67 percent solids and 99.5 percent sucrose. The product can be used wherever a dissolved granulated sugar product might be used. All liquid products are sold at the highest concentration of solids (sugar content) at which the sugar will remain in solution at normal room temperature (about 70°F). Very high-quality liquid sucrose with lower color and percentage ash are available from certain refiners.

Amber liquid sucrose is darker in color and higher in ash (inorganics) than liquid sucrose. It can be used where color and minor inorganic impurities are not critical.

Liquid invert, in which half of the sucrose present has been inverted (converted to a mixture of glucose and fructose), was developed mainly for the carbonated beverage industry. Partial inversion increases the solubility of the sugars in liquid invert, and allows these solutions to be stable at a level of 76 percent solids. At this higher level of solids, microbiological activity is less. It is also known that the mixture of these three carbohydrates, in the proper ratio, has an increased level of sweetness. This product only lends itself to uses in liquid products because the invert content retards crystallization of the sucrose. Liquid invert syrup of various refining purification are available.

Total invert syrup (Table 3-2) is the least-used liquid sugar. In this particular syrup, at least 93 percent of the sucrose has been converted to glucose plus fructose. This syrup is used mainly to retard crystallization of sucrose in food products.

Brown Sugars

Brown sugars are produced in a manner such that some of the sugarcane nonsugars are concentrated to impart a pleasant flavor. The higher the impurities level (percent ash, percent acids, etc.), the stronger the flavor. Brown sugars are used in the home and food industry to develop the rich molasses-type flavor in cookies, candies, and similar products.

There are three categories of brown sugars produced today. The original brown sugars are called soft brown sugars and are crystallized directly from the dark syrups obtained during the refining process. This process affords a very fine crystalline structure and a soft texture. The two major types of soft brown sugar are light (golden) and dark (Table 3-3). Some refiners will produce various grades within the light and dark range. They are normally numbered from 6 to 13, with 13 being the darkest. Color data shown in Table 3-3 are reported in terms of International Color Units (ICU). In selecting a brown sugar, one must evaluate the smooth flavor and color desired in the finished product.

Coated brown sugars are similar to soft browns except that they are produced by coating very fine granulated sugar crystals with a cane sugar syrup (Table 3-2).

Free-flowing brown sugars are a specialty product made by some refiners. In this case, the true crystals are an extremely fine powder produced by a cocrystallization method whereby the syrup holds the microcrystals together. These products have less moisture and are as free-flowing as granulated sugar.

Brown sugars are sold in 50- and 100-pound bags. The storage conditions of browns should be carefully observed to avoid lumping of the product. The product should be stored in controlled humidity areas to avoid drying out. Also, the products should be stacked in a manner to avoid compacting.

Specialty Sugars

Specialty sugars can be classified into two general categories: those serving the retail trade and those serving the industrial and institutional trade.

Specialty sugars can be defined as sugar products made to meet a specific need. In general, since they serve a small market, volume is limited and price is relatively high. In the retail market, specialty sugars can range from light and dark brown sugars to tablets and cubes. In industrial markets, the products can be even more selective. For example, some companies produce sugar products specially made for direct compaction or a fondant-quality sugar that can be extruded. Other sugar companies, recognizing the need for fiber in the diet, have isolated a natural fiber from sugar beet pulp for food formulators who wish to incorporate fiber in their products.

While some specialty products are well-known in the marketplace, others serving highly specific markets have limited visibility.

Molasses

The end product from sugar refining is molasses. Approximately 40 to 60 percent of molasses is sucrose and invert sugars, with the remainder being inorganic nonsugars. It is highly colored. The primary use of molasses is in animal feed supplements. Specialty molasses and syrups produced from cane juices are available for food flavoring.

Conclusion

As stated previously in this chapter, there are many uses for sugar. Various sugar products have been developed to meet product-specific requirements. Table 3-4 shows U.S. sugar deliveries to industrial and nonindustrial users from 1980 through the end of 1988.

Table 3-1. Granulated Sugar: Typical Analysis

	Coarse	Sanding	X-Fine	Fruit	Bakers Special	Pwd. Sugar 6X	Pwd. Sugar 10X
U.S. Screen Mesh							
% on U.S. 12	0–2	—	—	—	—	—	—
16	45–65	—	—	—	—	—	—
20	20–35	2–10	0–5	—	—	—	—
30	—	40–70	2–20	—	—	—	—
40	—	30–40	10–45	0–7	—	—	—
50	—	1–8	5–35	20–50	0–5	—	—
70	—	—	—	30–70	10–30	—	—
100	—	—	10–40	10–30	30–60	0–2	0–1
% Thru 100	—	—	0–8	0–10	10–30	—	—
% Thru 140	—	—	—	—	5–20	—	—
% Thru 200	—	—	—	—	—	88–100	94–100
Color (ICU)	20–35	20–35	25–50	25–50	25–50	25–50	25–50
% Ash (max)	0.015	0.015	0.02	0.03	0.03	0.03	0.03
% Moisture (max)	0.04	0.04	0.05	0.05	0.05	0.5	0.5
Starch	—	—	—	—	—	2.5–3.5	2.5–3.5

Table 3-2. Liquid Sugars: Typical Analyses

	Liquid Sucrose	Amber Sucrose	Liquid Invert	Total Invert
% Solids	67.0–67.9	67.0–67.7	76–77	71.5–73.5
% Invert	0.35 max	0.4 max	45–55	93 min
% Sucrose	99.5 min	99.4 min	55–45	7 max
% Glucose	—	—	23–28	46.5 min
% Fructose	—	—	23–28	46.5 min
Color (ICU)	35 max	200 max	35 max	40 max
% Ash	0.04 max	0.15 max	0.06 max	0.09 max
pH	6.7–8.5	6.5–8.5	4.5–5.5	3.5–4.5
Solids lbs/gal	7.42–7.55	7.42–7.52	8.75–8.91	8.05–8.35
Total wgt lbs/gal	11.08–11.12	11.08–11.11	11.52–11.57	11.25–11.36

Table 3-3. Brown Sugars: Typical Analysis

	Soft Brown		Coated Brown		Free-Flowing	
	Lt.	Dk.	Lt.	Dk.	Gran.	Pwd.
% Sucrose	85–93	85–93	90–96	90–96	91–96	91–96
% Invert	1.5–4.5	1.5–5	2–5	2–5	2–6	2–6
% Ash	1–2	1–2.5	0.3–1	0.3–1	1–2	1–2
% Organic non-sugars	2–4.5	2–4.5	1–3	1–3	0.5–1	0.5–1
% Moisture	2–3.5	2–3.5	1–2.5	1–2.5	0.4–0.9	0.4–0.9
Color (ICU)	3,000–6,000	7,000–11,000	3,000–6,000	7,000–11,000	6,000–8,000	6,000–8,000
Color (reflectance)	40–60	25–35	—	—	—	—
Granulation:						
% on U.S. 16	—	—	—	—	6% max	0
% thru U.S. 50	—	—	—	—	8% max	86% min
% thru U.S. 100	—	—	—	—	—	65% max

Table 3-4. U.S. Sugar Deliveries to Industrial and Nonindustrial Users

Type of User	Calendar Year								
	1980	1981	1982	1983	1984	1985	1986	1987	1988
	1,000 short tons, refined[1]								
Industrial users:	6,004	5,665	5,199	4,992	4,684	4,218	4,026	4,252	4,179
Food use	3,843	3,813	3,616	3,744	3,776	3,878	3760	4,040	3,942
Bakery and cereal products	1,337	1,306	1,296	1,387	1,404	1,494	1,432	1,513	1,541
Confectionery products	932	983	940	1,087	1,115	1,059	1,051	1,146	1,107
Dairy products	450	459	404	385	408	456	447	449	411
Processed foods	535	484	450	454	433	428	387	398	354
Other	589	581	526	431	416	441	443	534	529
Beverage use	2,161	1,852	1,583	1,248	908	340	266	212	237
Nonindustrial users:	3,353	3,421	3,214	3,076	3,053	3,123	3,075	3,199	3,316
Institutions	303	259	177	195	209	204	142	163	175
Eating and drinking	96	90	85	94	108	85	84	91	89
Other[2]	207	169	92	101	101	119	58	72	86
Wholesalers and retailers	3,050	3,162	3,037	2,881	2,844	2,919	2,933	3,036	3,141
Wholesalers, jobbers, and sugar dealers	1,881	2,001	1,951	1,713	1,744	1,874	1,867	2,040	2,200
Retail grocers, chain stores, and supermarkets	1,169	1,161	1,086	1,168	1,100	1,045	1,066	996	941
Total food and beverage use	9,357	9,086	8,413	8,068	7,737	7,341	7,101	7,451	7,495
Total other use[3]	120	126	106	131	127	131	138	149	121
All uses, continental U.S.	9,477	9,212	8,519	8,199	7,864	7,472	7,239	7,600	7,616

[1]Excludes Hawaii [2]Includes deliveries to government agencies and the military [3]Used largely for pharmaceuticals and some tobacco

Source: National Agricultural Statistics Service, USDA

References

1. Association of Official Analytical Chemists. *Official Methods of Analysis*. 15th ed. Association of Official Analytical Chemists, Arlington, VA. 1990.
2. National Food Processors Assoc. *Laboratory Manual for Food Canners and Processors, Vol. 1*. 3rd ed. AVI/Van Nostrand Reinhold, New York. 1968.
3. National Soft Drink Assoc. *Quality Specifications and Test Procedures for Bottlers' Granulated and Liquid Sugar*. National Soft Drink Assoc., Washington, DC. 1975.
4. Pancoast, H.M. and Junk, W.R., *Handbook of Sugars*, 2nd edition. AVI/Van Nostrand Reinhold. New York. 1980.
5. Schneider, F. *Sugar Analyses, ICUMSA Methods*. International Commission for Uniform Methods of Sugar Analysis, Peterborough, England. 1979.
6. Schultz, H.W., Cain, R.F., and Wrolstad, R.W. *Symposium on Foods, Carbohydrates and Their Roles*. AVI/Van Nostrand Reinhold, New York. 1969.
7. *The United States Pharmacopeia*. The National Formulary, Rockville, MD. 1980.

4

Properties of Sugar

R. L. Knecht*

This chapter provides an overview of the chemistry of sugar including some of its more important physical, chemical, and biological properties.

Chemistry of Sucrose

Cane and beet sugar are identified in technical terms as sucrose. In some parts of the world, sucrose is called saccharose, a word derived from the family name for all sugar compounds—saccharides. Sucrose is a natural product of green plants, which combine carbon dioxide, water, and the energy of the sun, in the presence of chlorophyll, to provide a means of storing energy. This combination of carbon dioxide and water makes sucrose a member of the carbohydrate family of organic compounds. In fact, sucrose is one of the most abundant carbohydrates found in nature and is a major component of the food chain.

The sucrose molecule is composed of twelve atoms of carbon, twenty-two atoms of hydrogen, and eleven atoms of oxygen and is written in chemistry shorthand as:

$$C_{12}H_{22}O_{11}$$

The molecular weight of sucrose is 342.30.

*R. L. Knecht is Senior Vice President, Operations, C. & H. Sugar Company, Crockett, CA.

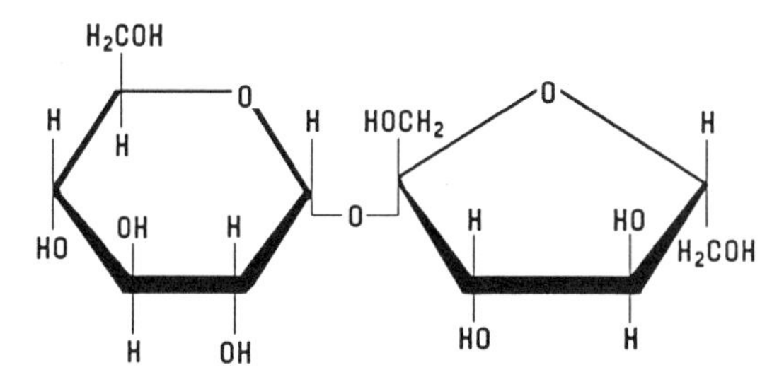

Figure 4-1. Chemical structure of sucrose molecule.

A more descriptive way of displaying the sucrose molecule is to show the manner in which the atoms are attached or bonded (Figure 4-1).

In chemistry, there exists a standard method of naming all chemical compounds in such a way that the appropriate structure can be drawn from that name. The International Union of Pure and Applied Chemistry (IUPAC) nomenclature for sucrose is:

α–D–Glucopyranosyl–β–D–fructofuranoside

We shall use the more common chemical name, sucrose, in place of the above.

The chemistry of carbohydrates is complex and includes all compounds that have a carbon base and have hydrogen and oxygen atoms present in a one-to-two-to-one ratio. A simple sugar is a carbohydrate that cannot be hyrolyzed into a simpler compound (that is, undergo a chemical reaction in which there is a net gain of two atoms of hydrogen and one atom of oxygen). These simplest of sugars are called monosaccharides. The common chemical names for all sugars end in "-ose."

Two important monosaccharides are glucose (dextrose) and fructose (levulose). These two sugars are also called blood sugar and fruit sugar, respectively. Sucrose is a disaccharide composed of one molecule of glucose linked to one molecule of fructose. Other disaccharides commonly found in nature include maltose (glucose-glucose) and lactose (glucose-galactose). Sugars can consist of hundreds to thousands of linked monosaccharides. When three to ten monosaccharides are linked, this sugar is called an oligosaccharide. When over ten simple sugars are linked, the resultant polymers are called polysaccharides. The most common examples of polysaccharides are starch and cellulose. Each consists of thousands of linked glucose molecules. Cellulose and starch exhibit different properties because the glucose monosaccharides are bonded together in different fashions in these polysaccharides.

As a disaccharide, the sucrose molecule can be split into the two component monosaccharides. The reaction that converts one molecule of sucrose into one molecule of glucose and one molecule of fructose is a simple hydrolysis reaction commonly called the inversion reaction. The resulting mixture of glucose and fructose is called invert sugar. Honey is a prime example of naturally occurring invert sugar. The inversion reaction is accelerated by the presence of acids (low pH) or by some enzymes, the most notable being invertase.

Physical Properties of Sucrose

Sucrose is normally used as a solid, as it is extremely stable in its crystalline form. Pure sucrose is colorless, odorless, and obviously sweet tasting. The crystalline form is termed monoclinic in the system for classifying crystals. Crystalline sucrose melts or decomposes between 160° and 186°C, the exact temperature depending on the solvent of crystallization and purity. The specific gravity of a single sucrose crystal is 1.588. In a large grouping of crystals, such as a pile of granulated sugar, the specific gravity will vary somewhat, depending on the average crystal size and size distribution along with the degree of compaction. The range of this variation is quite narrow, around 0.80 specific gravity. This translates to a bulk density of between 48 and 54 pounds per cubic foot (about 0.9 metric tons per cubic meter) for most products.

Solubility

Sucrose is extremely soluble in water and is also soluble in alcohol and other polar solvents. It is generally insoluble in ether, benzene, and other nonpolar organic solvents. The solubility of sucrose in water is one of its most studied properties since it is such an important factor in its production and use. Numerous tables and equations are available to determine its saturation point (maximum solubility) at various temperatures and conditions. One such equation is that of D. F. Charles, which expresses the solubility (S) as percent by weight in water as:

$$S = 64.397 + 0.07251t + 0.0020569t^2 - 9.035 \times 10^{-6}t^3$$

where t = degrees Celsius.

In more general terms, at the freezing point of water, approximately 180 grams of sucrose are soluble in 100 grams of water, and nearly 500 grams are soluble in water at its boiling point. By comparison, about 138 grams are soluble in 100 grams of alcohol (ethanol) at room temperature. In Table 4-5, a listing of some selected properties of sucrose/water solutions can be found.

Brix

Since the amount of sucrose dissolved in water is important industrially, a whole series of measuring indices were developed to indicate the relative proportions of the two materials. The most important of these is the Brix scale, which relates the percentage by weight of sucrose in a water solution. Therefore, 65° Brix would represent a solution that is 65 percent sucrose and 35 percent water. The Brix scale is also used to measure solutions other than pure sucrose and water. Thus, a Brix reading will normally be used to obtain the corresponding specific gravity or refractive index of a solution that, by practice, is agreed to correspond to a solution of pure sucrose and water at a set reference temperature, usually 20°C. A lesser-used index is the Baumé scale, which is another specific gravity ratio. With pure sucrose, the specific gravity or the refractive index of a water solution can be used to represent exactly the weight ratio of sucrose to water in that solution. With solutions that contain substances other than pure water and sucrose, these indices represent "apparent solids" or "apparent density," which may differ from the "true solids" depending on the type and amount of impurities.

Refractive Index and Specific Gravity

As noted above, the measurement of sucrose in solution is often expressed in terms of specific gravity, the ratio of the weight of a volume of solution to a similar volume of water. Sucrose solutions will also refract light in proportion to the amount of sucrose in solution, and this refraction is used as a measure of the solution density. The measure of this refraction is called the refractive index.

Since this refraction also varies with temperature and the wavelength of the light source, these two variables are normally specified when the refractive index is reported. They are usually determined at 20°C and at a wavelength that corresponds to the sodium D-line. The refractive index for a 100 percent sucrose solution is therefore expressed as:

$$N_D^{20} = 1.33299$$

The instruments used to measure this variable are called refractometers.

Optical Rotation

Like many organic compounds, sucrose is optically active, meaning that it will rotate the plane of polarization of a beam of polarized light passing through a solution. This is an important property for analytical procedures since the degrees of rotation of the beam of light will be proportional to the amount of optically active material present.

As with the refractive properties discussed above, optical rotation is also dependent on temperature and wavelength. It is also dependent on cell length (the distance the light passes through the solution). By holding each of these three variables constant, it is possible to measure the amount of sucrose present in a pure solution. The property of optical activity is normally defined as specific rotation and abbreviated as α. By convention, if a 100-millimeter cell length, a temperature of 20°C, and a wavelength of light corresponding to the sodium D-line (589 nanometers) are used, the specific rotation of sucrose is:

$$\text{Sucrose } [\alpha]_D^{20} = +66.53°$$

The positive sign indicates that the rotation of the plane of polarization will be to the right, or dextrorotatory. When sucrose is broken down into its two component monosaccharides, the resulting solution will rotate the same polarized beam to the left, or levorotatory. This inversion from dextro- to levorotation has given the name invert sugar to the fructose-glucose mixture. Since the glucose component is dextrorotatory like sucrose, it acquired the common name dextrose. The fructose molecule is levorotatory and is, therefore, often referred to as levulose. Since fructose has a larger negative specific rotation than dextrose's positive specific rotation, the mixture of 50 percent of each (invert sugar) yields a left rotation.

$$\text{Glucose } [\alpha]_D^{20} = +52.7°$$

$$\text{Fructose } [\alpha]_D^{20} = -92.4°$$

A separate scale is used in the sugar industry to define the specific rotation of sucrose. A standard pure sucrose solution is defined as 100°S and a pure water solution as 0°S. The instrument used to measure optical rotation is called a polarimeter, while one calibrated in degrees S is more properly called a saccharimeter. Since this scale is so common in sugar analysis, often the term "pol" is used to refer to the measured rotation on a saccharimeter. If pure sucrose is dissolved in water, the pol will measure the sucrose in solution. As with other analyses, this method is also applied to impure solutions containing other optically active molecules. With nearly pure solutions, this methodology serves well to approximate the sucrose content. When the sucrose content, expressed as pol, is divided by solids content, expressed as Brix, the result is multiplied by 100 and termed the apparent purity. True sucrose divided by actual solids is termed true purity. The lower the sucrose purity, the larger the discrepancy between apparent and true purities.

In 1986, the International Commission for Uniform Methods of Sugar Analysis (ICUMSA) approved a revised method for measuring the optical rotation of a pure sucrose solution. This revision was based on new technologies in equipment and procedures and afforded a more precise measurement of optical rotation. With this procedure, the wavelength of light used to measure optical rotation corresponds to the mercury green line, or 546.227 nanometers. For commercial purposes this change is trivial, but, to avoid confusion, when this scale is used to measure the optical rotation of sucrose, the result is reported in terms of degrees Z. Since this change is related to the redefinition of the 100 degree point, it is easy to convert from one scale to the other. Degrees S times 0.99971 will equal degrees Z. For most applications outside the laboratory, the two scales can be considered identical since the change represents a shift of only 0.029 percent.

Viscosity

An important functionality measurement of any fluid is its viscosity, or its resistance to flow. The reciprocal of viscosity is fluidity. These two terms are important in many applications, including finished product and engineering considerations. The coefficient of viscosity is meaured in dynes, and this term corresponds to the measured flow of a liquid under specified conditions. The dyne measurement is the force required per unit area to maintain a set velocity of flow. A more convenient and common measurement is the Poise, which relates the coefficient of viscosity to that of water.

In sucrose solutions, viscosity increases with solids content, although not in a linear fashion. Viscosity also decreases rapidly with an increase in temperature. And, as a general statement, as sucrose purity decreases its viscosity will increase.

Full expression of this relationship for pure sucrose solutions requires tables relating viscosity to temperature and concentration. At 20°C, a 20° Brix solution of sucrose has a viscosity of about 2 centipoise, a 40° Brix solution will be about 6 centipoise, a 60° Brix solution about 60 centipoise, and a 70° Brix solution about 480 centipoise. Heating a 60° Brix sugar solution from 20°C to 50°C will lower its viscosity from 60 to 12 centipoise.

Specific Heat

This property measures the energy required, in calories, to raise the temperature of one gram one degree Celsius. In the English system the same property is measured as the BTUs required to raise one pound one degree Fahrenheit. The specific heat for sucrose is 0.63 calories. Since sugar is often used in solution, the specific heat of the water-and-sucrose combination is an important property. At 20°C, the specific heat of a 60° Brix pure sucrose solution is 0.66 calories. One expression for the specific heat of sucrose and water solutions is:

$$S = 1 - [0.632 - 0.001T + 0.0011(100 - P)] \times \frac{B}{100}$$

where S = Specific Heat

T = Temperature in degrees Celsius

P = Purity (pure sucrose = 100)

B = ° Brix.

Heat of Solution

When crystalline sucrose is dissolved in water, the temperature of the solution decreases. This is because sucrose has a negative heat of solution of about 2 kilocalories per mole. (A mole of sucrose is 342.30 grams.) Interestingly, for amorphous or milled sucrose, the heat of solution is positive, at about 3.5 kilocalories per mole. A decrease in temperature is also observed on diluting a solution.

Heat of Crystallization

As sucrose crystallizes, it gives off heat. This is termed the heat of crystallization, and it is approximately 2.5 kilocalories per mole at 30°C to 8 kilocalories per mole at 60°C.

Boiling Point Rise, Freezing Point Depression

Sugar in solution has the effect of lowering the freezing point of that solution while raising the boiling point. In many applications this is an important effect. The relative change in freezing or boiling point is proportional to the amount of sucrose in solution. This effect is expressed as:

at atmospheric pressure, 1 mole of sucrose dissolved in 1 kilogram of water will depress the freezing point by about 2° Celsius and,

at atmospheric pressure, 1 mole of sucrose dissolved in 1 kilogram of water will raise the boiling point by approximately 0.5° Celsius.

Since molor concentrations of sugar are not used in most applications, these physical properties of sucrose are normally described in terms of Brix (Table 4-1).

Table 4-1. Effect of Sucrose Concentration on Freezing and Boiling Points of Water

°Brix	Freezing Point Depression (°C)	Boiling Point Elevation (°C)
10	0.61	0.15
15	1.01	0.25
20	1.50	0.40
30	2.6	0.70
40	4.5	1.20
50	—	2
60	—	3
70	—	5

Surface Tension

This property of sucrose causes an increase in surface tension as the amount of sucrose in solution increases. This relationship at room temperature is shown in Table 4-2.

Table 4-2. Effect of Sucrose Concentration on Surface Tension of Water

°Brix	Surface Tension (dyne/cm^2)
0 (pure water)	72.7
10	73.4
20	75
40	77
60	79

Angle of Repose

Bulk crystalline white sugar will assume a natural angle of repose between 30° and 37°. For most purposes, a 34° angle of repose can be assumed. This number equates approximately to the angle of slide.

Vapor Pressure

The vapor pressure of water in an aqueous sucrose solution will decrease with an increase in dissolved sucrose. As measured in millimeters of mercury, the vapor pressure of water will vary as shown in Table 4-3.

Table 4-3. Effect of Sucrose Concentration on Vapor Pressure of Water

°Brix	40°C	80°C
0	55.3	355.2
10	54.7	354.0
20	54.3	352.3
60	52.0	320.4
70	51.1	298.4

Water Activity/Equilibrium Relative Humidity

Another important property of sucrose is its ability to "tie up" water. This property affects the water activity (a_w) of the solution or product in which sucrose is present, thereby impacting, among others, its look, texture, mouthfeel, and keeping or shelf-life. Water activity is defined as:

$$a_w = \frac{P}{P_o}$$

where P = partial vapor pressure of food moisture, and

P_o = saturation vapor pressure of water.

Water activity is sometimes referred to as ERH or Equilibrium Relative Humidity. Sucrose allows a much lower a_w for the same moisture content, and this, in effect, retards the rates of many reactions, including microbiological spoilage. The impact of sucrose on water activity in foods is a very complex relationship and is not easily expressed. One comparison of this ability to tie up water: a solution of 44° Brix sucrose (56 percent moisture) has the same water activity, 0.8, as a mixture of starch and water with a moisture of 20 percent. This is why one major use of sucrose has been in preservation of foods.

Electrical Properties

Sucrose is a nonconductor. A pure solution of sucrose will have no conductivity as measured against a water standard. Sucrose will also reduce the conductivity of salt solutions. Pure sucrose, when dissolved, will form a neutral solution (neither acid nor alkaline; 7.0 pH). Commercial sugar will normally have a pH of 7.0 to 7.15 in solution.

Osmotic Pressure

Osmosis refers to the diffusion of a substance through a semipermeable membrane and is important in biological applications. The osmotic pressure of an aqueous sucrose solution has been shown to vary with both concentration and temperature. Some reported osmotic pressures, in kilograms per centimeter at various Brix and temperatures, are shown in Table 4-4.

Table 4-4. Effect of Sucrose Concentration and Temperature on Osmotic Pressure

	Temperature		
°Brix	0°C	30°C	60°C
5	4	4.2	4.6
10	8.8	9.3	10.1
10	18.8	21.2	21.7
40	55	57.5	—

Chemical Properties of Sucrose

All sugars are fairly reactive. However, with sucrose, the main reactive sites of the glucose and fructose molecules are bonded together when the sucrose molecule is formed. The two monosaccharides are therefore much more reactive than sucrose. In fact, they are termed reducing sugars because of their ability to act as chemical reducing agents. Sucrose is a nonreducing sugar and is stable in heated, neutral solutions up to 100°C. Fructose, however, will decompose above 60°C, and both glucose and fructose are unstable in alkaline solutions, conditions under which sucrose is most stable. In acidic solutions, sucrose will invert or break down into its two component monosaccharides, glucose and fructose. This reaction will accelerate with increasing acidity and increasing temperature. Most sucrose reactions in solution, including human metabolism, begin with the inversion reaction.

1. *Inversion Reaction.* This is an irreversible hydrolysis reaction in which one sucrose molecule and one water molecule yield one molecule of glucose and one molecule of fructose. As above, this reaction proceeds more quickly with heat. Inversion of a pure sucrose solution proceeds nearly 5000 times faster at 90°C than at 20°C. In practical terms, this reaction takes place below pH 7 and proceeds more quickly as the pH decreases. The reaction is endothermic with an activation energy of 25.9 kilocalories per mole at 20°C. This reaction can also proceed by biochemical catalysis with several enzymes, most notably invertase.
2. *Thermal Decomposition.* Dry, crystalline sucrose is fairly stable up to its melting point. Molten sucrose will decompose

rapidly into glucose and fructosan. If it is heated to around 200°C, a whole series of decomposition products are formed. A nonsweet, dark brown, water-soluble mixture called caramel is produced. Further heating will eventually lead to a carbon residue. In the presence of air or oxygen, this thermal decomposition becomes combustion, forming eventually carbon dioxide and water. The energy released by full combustion of sucrose is 3.95 calories per gram.

3. *Maillard Reactions.* The reducing sugars formed from sucrose will react with amino acids and proteins. This is one of the more important sets of reactions in food preparation and yields a family of brown-colored products, known as melanoidins, along with volatile compounds with strong odor and flavor intensities. These reactions are the basis for the browning and aroma-flavor formations associated with heating foods.
4. *Acidic Degradation.* In the most general sense, in the presence of acids, especially at a pH below 3, the resulting monosaccharides from the inversion reaction will form various condensation reaction products. The result is a series of oligosaccharides such as isomaltose and gentiobiose along with furan-type derivatives.
5. *Alkaline Degradation.* Again, in general terms, the alkaline degradation products of sucrose are more likely to include cleavage products such as organic acids. However, once this "brew" starts, the primary products participate in further reactions such as aldol condensations and the formation of cyclic compounds. Ketones and acids are the major groups formed in these additional reactions.

Biological Properties of Sucrose

Sucrose is an important feed stock for many critical biological reactions, the most important being human metabolism covered in Chapter 7. Since sugar is the product green plants form during conversion of sunlight into storable food energy, it is not an exaggeration to say that the reconversion of this stored energy to usable energy in living things is perhaps the most important biological reaction in the food chain. Other than the role of sugar in human metabolism, there are many other critical biological reactions that are well documented in the literature. Some of the most noteworthy are described below.

1. *Yeast Fermentation.* Whether it is the action of yeast in causing bread to rise or in converting grape juice into wine, the reaction is that of yeast fermentation. This reaction produces alcohol and carbon dioxide gas. In baking, a large portion of the sugar added to the product is converted into volatile compounds that are given off as gas and vapors in the oven. The sugar not converted will react chemically as described on page 57. (Cooking is really the art of controlling chemical reactions in foods.)

   In alcoholic beverage production, this same reaction occurs although different strains of yeast are used. Wine making converts the 18- to 24-percent sucrose content of the pressed grape juice into 10 to 13 percent alcohol by volume. It is the variable content of sugar and other juice components, along with the variable controls the winemaker employs to modify these reactions, that make the production and appreciation of wines such a complex and interesting field. Sweet wines and dessert wines are made by adding additional sugar to the fermenting mixture to keep the reaction going.

   In some areas, most notably Brazil, this same reaction is used to produce power alcohol to replace a large percentage of the gasoline used to power internal combustion engines.

2. *Leuconostoc Bacteria.* This strain of bacteria is important because it produces dextran, a long-chain polysaccharide, as a product. While dextran has some important commercial uses (such as a blood plasma extender), this product is a sanitation problem in most operations as the gelatinous consistency and the viscosity impact of these products are undesirable.

3. *Biotechnology.* Lactic acid, butyric acid, acetic acid, citric acid, levan, ketose, and many other chemicals are commercially produced by biochemical means with sucrose used as a substrate or starting material.

In most commercial biochemical procedures, reactions occur with sucrose in aqueous media. In solutions where water activity is below 0.4, most biological reactions will not proceed. Therefore, high Brix solutions tend to reduce the rate of these reactions. Similarly, biological reactions can be inhibited with either very low or very high pH values, or extremes in temperature. In fact, temperatures above 180°F will generally kill or immobolize all but the thermophilic bacteria. Bacterial spores, however, can survive even these extremes and reactivate when conditions moderate. Therefore, low Brix, warm temperatures, and moderate pH levels are optimum conditions for most biological activity.

Other Sugar Products

The discussions above dealt with pure sucrose and its properties as a solid and in an aqueous solution. The physical properties of other significant sucrose-containing products may vary widely from those discussed previously while chemical properties are generally less variable.

1. *Brown Sugars (Soft Sugars)*. These products are predominantly sucrose (85 to 95 percent) with a notable amount of invert sugars (1 to 4 percent) and moisture (1 to 3 percent). Present to a lesser extent will be some inorganic salts and sucrose by-products such as caramels. The moisture and amount of free syrup in the brown sugars dictate different handling from the free-flowing crystals of pure sucrose.

2. *Powdered Sugars*. These, as the name implies, are pulverized crystals of sucrose, usually with 3 percent corn starch added to reduce caking. The corn starch is only slowly soluble in water.

3. *Liquid Sugars*. These sugars are usually produced as pure sucrose solutions or solutions of sucrose with varying levels of invert sugar. Another series of products often called refiners syrups or liquid browns are somewhat analogous to dissolved brown sugars.

4. *Molasses*. Refiners molasses (blackstrap molasses) is an inedible concentration of the nonsucrose materials from the refining process. However, some edible molasses products are produced. These products are normally similar to a higher invert content or higher Brix version of the refiners syrups noted above.

5. *Agglomerated Sugars*. These are specially prepared products with a microcrystalline structure to confer tailored physical properties and special functionalities. Products in this group are used in glaze formations, fillings, and tableting operations, for example.

Storage and Handling of Sugar Products

Commercially distributed sugar products are available in a variety of forms and packages. Packages of white granulated sugar products range from packets containing one-tenth of an ounce up to bulk rail cars of 200,000 pounds. Because of the stability and unique handling properties of sucrose, long-term storage of these products, even in the most inexpensive packaging configurations, can be accomplished without product degradation under normal storage conditions.

Over the years, sugar packaging has moved from the barrel to the cotton sack to, currently, paper and plastic as the normal commercial packaging materials. Multiwall paper bags of various granulated, powdered, brown, and specialty sugar products are generally available in sizes from 25 to 100 pounds. Consumer-sized packages are also available in single or multiwall bags of 2, 5, and 10 pounds. These smaller bags are normally baled or bundled into a 60-pound shipping unit. One- and 2-pound cartons of granulated, brown, and powdered sugar are available and are normally cased in a corrugated shipping container with a net weight of 24 pounds. More recently, single-ply polyethylene 2- and 4-pound packages of brown and powdered sugars have become available. These bags are baled or cased as a 24- or 32-pound net shipping container.

Granulated Sugar Products

Granulated sugar products are basically masses of single sugar crystals, and are differentiated by the size distribution of the crystals. The sugar refining process produces these crystals to be free-flowing by conditioning the sugar to remove almost all the free moisture. The total moisture in most granulated products is 0.03 percent. There is not enough moisture present to cause lumping or caking even after prolonged storage. External moisture, however, can be absorbed by the sugar. Therefore, it is important that sugar products be not exposed to storage conditions that will promote this absorption of external moisture.

Moisture migration is promoted by temperature gradients, the difference between the temperature and the relative humidity of the product and those of the surrounding environment. Therefore, it is important to store granulated products in relatively dry environments that are not exposed to rapid changes in temperature. This will minimize the chances of moisture pickup and possible caking.

Properly cured sugar that is packaged at a sugar temperature below 105°F will not cake or form lumps in transit or storage. If stored under normal warehousing conditions (40 to 100°F, 40 to 70 percent relative humidity) and not exposed to rapid fluctuations, sugar will remain stable and free-flowing for years. The shelf-life of sugar is a function of the storage conditions, with the only major problem encountered being lumping. As an extremely pure crystal, it is very stable and is not prone to decomposition or loss of quality.

Granulated sugars have an angle of repose of approximately 34 degrees. These products are easily conveyed on scroll or belt conveyors. The crystalline nature of the product is also well suited to pneumatic or vacuum transport. Obviously, with such a free-flowing product, gravity distribution via chutes or pipes is often used.

Crystalline sugar is noncorrosive but, as with any moving solid, is abrasive. Like all carbohydrates, sugar dust in air can be a hazard. Sugar is combustible, and care must be exercised in its handling to prevent the ignition of a mixture of air and sugar dust. Under certain conditions, this mixture can be explosive if supplied with an external ignition source.

Liquid Sugars

Liquid products are solutions of sugar and water. As a solution, the product can be stored and transported in tankage and, in some cases, in drums or pails. Most of these products are produced with levels of invert sugar varying from 0 to 100 percent. In almost all cases, the product is adjusted to a specified pH as part of the final processing. Liquid sugars are sold on a Brix or equivalent solids basis, with most products normally in the 65 to 75 percent solids range.

Liquid products need to be protected from microbiological attack. Almost all sugar products are excellent substrates for mold, yeast, and bacterial growth. Therefore, sanitation and sterilization of tankage, pumps, and transport and contact surfaces are important. At the relatively high Brix of production, shipping, and storage, growth of most microorganisms should be controllable. It is almost impossible, however, to prevent a thin layer of dilute product accumulating on the surface from the condensation of the vapor in the headspace of an enclosed storage vessel. This thin film of low Brix material is generally warm and a perfect medium for inoculation and growth of microbiological contamination. Two means of reducing this exposure are a sterile, filtered air-sweep of the headspace to remove vapors and germicidal ultraviolet lamps to keep the headspace and the surface film sterile. Smooth, easily cleaned and sanitized surfaces are also generally engineered into liquid-sugar systems.

The loss of water from the solution to evaporation, or the chilling of these concentrated solutions, can lead to crystals of sucrose forming and dropping out of solution. Usually, the addition of water, heat, or circulation can reconstitute the product easily. Heating the solution or lowering the Brix will also decrease the viscosity of the product and make pumping easier.

In solution, sucrose is not in its most stable crystalline state and will therefore be more prone to loss of quality. Inversion and color formation are generally noted in prolonged storage. Liquid products are surprisingly stable, however, and under proper storage conditions can be maintained for periods of weeks prior to use.

Powdered Sugars (Pulverized Sugars)

Sugar crystals can be fractured and ground into an amorphous mixture of fine particles. As such, the resulting product has a very large surface area-to-weight ratio. The properties of powdered sugar products require unique methods of storage and handling. These products are compressible, and their large surface area allows them to pick up moisture easily. This makes powdered sugars especially prone to lumping and caking.

While granulated products can be stored with little concern for pressure set, powdered products are rarely stored more than two pallets high in a warehouse. Also, the fine particles that make up this product are more prone to dusting than their granulated counterparts; therefore, dust control is an important parameter in handling this product. Shelf-life of powdered sugar products is decidedly shorter than for granulated products because of this lumping tendency. Even so, under proper storage and handling conditions, powdered sugar is stable for long periods. If an anticaking agent is added during the production process, powdered products can be stored for months with no loss of quality.

Brown Sugars (Soft Sugars)

Brown sugars are a mixture of crystals and syrup. Moisture loss is the main storage problem with these products. If enough moisture is allowed to evaporate from the syrup film on the crystals, the syrup itself will begin to thicken or crystallize and cement the mixture together. In extremely cold storage, the viscosity of this film will increase enough to give the appearance of hardening. For these reasons, brown products are normally packaged with moisture barriers to reduce this moisture loss. Even so, prolonged storage under dry and hot conditions can lead to the product turning hard. Unless these conditions are extreme, the shelf-life of brown sugars can be measured in months, but under proper conditions these products will be very stable for longer periods.

The very mixture of crystals and syrup that gives brown sugars their unique flavor makes handling them markedly different from handling granulated products. Brown sugars will tend to be sticky and are not free-flowing bulk solids. They are compressible products and are generally conveyed by means of scroll conveyors. The syrup film makes them more reactive chemically and biologically; however, the dusting problem is reduced to insignificance.

Other Sweeteners

Obviously, sucrose is not the only food ingredient that is sweet. In Table 4-6, sucrose is compared to other nutritive and nonnutritive sweeteners that are in general usage.

Table 4-5. Selected Properties of Pure Sucrose/Water Solutions (All readings at 20° unless otherwise noted.)

Brix°	Refractive Index (η_0^{20})	lbs solid/ gallon	Total weight/ gallon	Specific Gravity (20/20)	Boiling Point (°C at 760mm Hg)	at 760mm Hg)	Specific Heat	Baumé (Modulus 145)
0	1.33299	0.0	8.322 lbs	1.00000	100.0	—	1.00	0
5	1.34027	0.424	8.485	1.01965	100.1	—	—	2.79
10	1.34783	0.866	8.655	1.03998	100.15	—	0.94	5.57
15	1.35567	1.325	8.830	1.06104	100.2	—	—	8.34
20	1.36384	1.802	9.012	1.08287	100.3	1.957	0.88	11.10
25	1.37230	2.300	9.201	1.10551	100.5	2.463	—	13.84
30	1.38108	2.819	9.396	1.12898	100.6	3.208	0.83	16.57
35	1.39017	3.368	9.599	1.15331	100.8	4.352	—	19.28
40	1.39968	3.924	9.809	1.17853	101.1	6.201	0.78	21.97
45	1.40962	4.512	10.027	1.20467	101.4	9.449	—	24.63
50	1.42002	5.126	10.252	1.23174	101.8	15.54	0.72	27.28
55	1.43079	5.767	10.486	1.25976	102.4	28.28	—	29.90
60	1.44188	6.436	10.727	1.28873	103.0	58.93	0.66	32.49
65	1.45342	7.135	10.977	1.31866	103.9	148.2	0.64	35.04
70	1.46541	7.864	11.234	1.34956	105.0	485.0	0.61	37.56
75	1.47784	8.624	11.018	1.38141	106.5	2344	—	40.03
80	1.49060	9.418	11.773	1.41421	109.4	—	0.55	42.47
85	1.50366	10.246	12.054	1.44794	113.2	—	—	44.86
90	—	11.108	12.342	1.48259	119.3	—	0.50	47.20
95	—	12.007	12.639	1.51814	128.9	—	—	49.49

Source: NBS Circular 44, C440; D. F. Charles, E. J. Culp Meade, "Cane Sugar Handbook"

Table 4-6. Comparative Sweeteners

Sweeteners	Chemical Formula	Relative Sweetness (10% Sucrose Soln. = 1.0)	Stability	Water Solubility (g/100g of soln. 25°C)	Chemical Type
Sucrose (Sugar)	$C_{12}H_{22}O_{11}$	1.0	Stability in neutral pH	67	Disaccharide (glucose + fructose)
Glucose (Dextrose)	$C_6H_{12}O_6$	0.6	Maillard Reactions	51	Monosaccharide, aldohexose
Fructose (Levulose)	$C_6H_{12}O_6$	1.2-1.8	Very Reactive	81	Monosaccharide, ketohexose
Invert	$C_6H_{12}O_6$	1.0	Very Reactive		Even mixture of glucose & fructose
Lactose (Milk Sugar)	$C_{12}H_{22}O_{11}$	0.3	Maillard Reactions	16	Disaccharide (galactose + glucose)
Sorbitol	$C_6H_{14}O_6$	0.6	Stable	72	Sugar alcohol (from glucose)
Mannitol	$C_6H_{14}O_6$	0.7	Stable	18	Sugar alcohol (from fructose)
Xylitol	$C_5H_{12}O_5$	1.0	Stable	64	Sugar alcohol (from xylose)
Saccharin	$C_7H_4O_3NSNa$	250-550	Stable	—	
Cyclamate	$C_6H_{12}NNaO_3S$	30-50	Stable		
Aspartame	$C_{14}H_{15}N_2O_5$	120-200	Unstable		Protein

Source: *Sweetness*, J. Dubbing, Editor, Springer-Verlag, 1987. *Carbohydrate Sweeteners in Food and Nutrition*. P. Koivistonen, Ed. Academic Press 1980

5

The Flavor of Sugar in Foods

R. L. Knecht*

Flavors and the perception of flavors are subjects of scientific complexity. Terms such as intensity, persistence, onset, threshold, and discrimination have importance in describing the perception of a flavor even before it is determined whether the flavor is liked or disliked. When sweetness is described in these terms, the standard of reference is sucrose. In recent years, there have been numerous attempts to study the sense of taste in these terms, but humankind has described preferences for certain flavors long before the study of taste became a science. It seems that nature has coded flavors to broadly describe the attributes of food. Sweetness generally indicates a food as desirable while bitterness leads to the opposite conclusion. This natural code is apparently imprinted and is not a learned preference. Newborn babies have been shown to have a preference for sweetness, even over milk. This same preference for sweetness is a trait humans share with all other mammals. Therefore, independent of the science of flavor and perception, some rather definitive conclusions on the reaction to sugars in the flavors of foods can be drawn.

Flavor of Sucrose

Sucrose plays a number of roles in food applications. It is rarely consumed alone. Instead, it imparts certain properties to the food mixture that affect its nutritional, physical, chemical, microbiolog-

*R. L. Knecht is Senior Vice President, Operations, C. & H. Sugar Company, Crockett, CA.

ical, and sensory characteristics. The wide functionality of sucrose makes it a unique flavorant.

As a multifunctional flavorant, sucrose has one other property that makes it unique—the flavor it imparts is pure sweetness without aftertaste or unpleasant secondary reactions. In most food combinations, the perception of individual flavor components is modified by all the others present. Some flavors depress perception of other components while others synergize the flavor sense. The relative purity of the flavor perception of sucrose complements its physical properties in controlling the variables of a multicomponent food product. Sucrose allows the addition of desired functional properties without introducing an unwanted flavor reaction.

Since sucrose is the standard by which the sweet flavor perception of all foods is measured, the attributes of that flavor are well understood under a whole series of variables, including temperature, concentration, acidity, color, consistency, mouthfeel, and the presence of other flavors and components. The sweetness of sugar is discussed in Chapter 6. In that chapter, the measurements of sweetness and the various permutations of the sweet taste perception are covered in some detail.

The words "taste" and "flavor" are often used interchangeably. To be precise, there are only four tastes humans detect: sweet, bitter, salt, and sour. Flavors result when these tastes are overlaid with the olfactory perception that occurs as a food is consumed. Coincidentally with taste and odor perceptions, the temperature, mouthfeel, and other characteristics of the food are sensed during consumption. All these senses occur with their own onset, intensity, and duration over a fairly short period of time. If the net result is pleasant, the food is said to taste good.

In this complex system, sucrose plays a major role that goes far beyond the sweetness perception. The properties of sugar add to or modify the perceptions of other food components in various ways. In order to describe these, sugar has been placed in the context of its major uses, and its role in each has been defined. In other chapters, the role of sugar in specific applications is described by specialized practitioners.

In complex food systems, sucrose has been shown to balance the flavor components. For example, the addition of small amounts of sucrose to corn, carrots, and peas produces a better-tasting product by minimizing the starchy and earthy notes while highlighting other desirable flavors. Fruit flavor studies have shown that the presence of sucrose in concentrations below 15 percent yields an increased perception of the desirable fruit flavors. In similar studies, sucrose and citric acid solutions tended to yield a perceived reduction in the sweetness of sucrose while at the same time reducing the perceived sourness of the citric acid. The opposite effect was noted in saline solutions. The presence of small amounts of sodium chloride enhanced the perception of sweetness from sucrose. The presence of small amounts of sucrose in a saline solution reduced the salty taste.

As noted, one important aspect of flavor perception is aroma. A number of projects have shown that the presence of sucrose in fruits and vegetables enhances the aroma of those food products. Sucrose therefore blends well with the flavors of fruits and vegetables and has been shown to increase the consumer's perception and acceptance of the flavors. At the same time, it enhances the aroma of the product while balancing the flavor notes. The well-understood chemical reactions of sugar that occur in cooking also produce a complete set of aromas that add materially to the flavor of cooked foods. In addition, sucrose helps to stabilize the color of a food and extend its shelf-life.

As the above indicates, individual flavor components can have several different functions in different food environments. These flavor functions are normally described as enhancement, suppression, and addition. Sugar has flavor functionality in all three of these areas. The complex perception of flavor has numerous facets, and the role of sucrose in modifying these perceptions generally involves the physical and chemical properties of sucrose to enhance or suppress the other components while providing the addition of pure sweet taste. The physical properties can lead to binding, releasing, or masking other flavors because of the changes imparted to the mixture's viscosity or volatility, for example. The chemical properties of sucrose can lead to direct reactions that change the chemical nature of individual components, or they can otherwise modify the system—for example, by changing the pH, and thereby changing the perceptive availability of other flavor components.

It is only recently that flavor chemists have begun to understand and describe these sensory perceptions in scientific terms. Sugar, on the other hand, has been studied in real applications and results for hundreds of years. We are now beginning to understand how and why grandma's apple pie tastes so good and sugar's role in our taste perception. The fact that sugar is a component of the total system that yields a pleasant sweet taste has been recognized for a long time.

Other Sugar Products

The sweet flavor described for pure sucrose is obviously an important component of the other sugar-based products. Brown sugars add a rich flavor described as "caramel" and "butterscotch" by most observers. These flavors are more intense and aromatic. Several other nuances are added by the various edible molasses products available. Powdered sugars, on the other hand, are most often used as a sweet carrier for other flavors in fondants and icings.

Numerous recent gas chromatography and mass spectroscopy studies have attempted to isolate and identify the flavor and aroma components of the brown sugar and molasses products. The studies have found a complex and varied mixture that has only begun to be unraveled. For a major food component whose use has been well understood for many years, the number of flavor constituents present is a graphic indication of the true complexity of flavor perception. Perhaps the best perception test available is still individual preference.

Flavor is a perception that consists of many varied components. Since these components all interact to yield the total perception, it is important that sugar be viewed in the complete context of the food system and not just as a sweet taste. Sugar is obviously a sweetening agent that imparts a clean and pure sweet taste to the food system. It is also a flavor synergist and modifier of other components, a texturizer, bulking agent, humectant, preservative, viscosity regulator, flavor substrate, and binder that also depresses the freezing point or raises the boiling point. In baked goods and doughs, it is a substrate for the chemical reactions of cooking or the biological reactions of fermentation. Sucrose brings all these well-understood functions to the food system, and each has an impact on the flavor of the finished food. Each of these functional and sensory properties has been well documented on a whole range of specific food applications. The study of the flavor of sugar is the study of the sum of all these parts in all their various applications. The very complexity of human sensory perception, and the chemistry of foods, make it impossible to describe flavor outside the full context of the food system.

References

For further details on the roles of sucrose in food flavor not contained in the chapters of this book, the following references provide an excellent starting point:

Lachmann, A., Dr. *The Role of Sucrose in Foods.* International Sugar Research Foundation, 1975.

Koivistoinen, P., and Hyvöner, L. (eds.). *Carbohydrate Sweeteners in Food and Nutrition.* Academic Press, London. 1980.

Dobbing, J. (ed.). *Sweetness.* Springer-Verlag, London. 1987.

Schultz, Cain, and Wrolstad. *Carbohydrates and Their Roles.* AVI/Van Nostrand Reinhold, New York. 1969.

Hornstein. *Flavor Chemistry—Advances in Chemistry 56.* American Chemical Society, Washington, DC. 1966.

6

The Sweetness of Sugar

Israel Ramirez*

Because nearly everyone knows what sucrose tastes like, the taste of sucrose would seem to be the most trivial of subjects. It is therefore surprising how much research has been done on the taste of sucrose and other sweeteners. Studies of sweet-tasting substances have been particularly important in psychology and food science. However, research on the taste of sugar has been undertaken by specialists in other disciplines as well. For example, biochemists and physiologists have been interested in receptors for sugars as well as how stimulation of the tongue affects neural processes. Unfortunately, vigorous research has led to a large number of isolated facts but few general principles. This chapter attempts to survey the most important facts and theories. The emphasis is on basic research that may be of use to food technologists.

Measurement

There are two main approaches to the measurement of sweet taste: psychophysics and psychometrics. Psychophysicists may be likened to engineers attempting to analyze the sensitivity of a sensor by examining the response to various concentrations of sweetener (14). Psychometricians get their inspiration from the paper-and-pencil techniques developed for measures of IQ and personality.

---

*Israel Ramirez is Assistant Member, Monell Chemical Senses Center, Philadelphia, PA. This work was supported by a National Science Foundation grant (BNS-8719309) and the Monell Chemical Senses Center.

There are two basic psychometric techniques: thresholds and magnitude estimation. Measurements of detection thresholds were popular in older studies. In this technique, subjects are given a series of solutions containing small amounts of sweetener or no sweetener and asked if they can detect the presence of some solute (absolute threshold=0.004 molar for sucrose) or sweet taste (recognition threshold=0.006 molar for sucrose [33]). This technique has fallen out of favor partly because it is laborious and partly because it emphasizes the least interesting concentrations of sucrose (highly dilute solutions). This technique could be applied to highly sweet solutions by asking the subjects which solution is sweeter than another ("just noticeable difference"), but the amount of labor involved has discouraged the use of this technique.

Magnitude estimation is relatively easy for experimenters to do. In this method, people are asked to produce a numerical response proportional to the stimulation received. The author's experience as a subject in such studies made him skeptical because it can be difficult for the subject to decide what number to assign to a given solution. For most sensory modalities, magnitude estimates of stimuli follow an exponential rule:

$$R = kC^{\beta}$$

where perceived intensity (R) is equal to a constant (k) times stimulus concentration (C) raised to some power (ß). Thus, logarithmic plots of magnitude estimates versus stimulus concentrations are supposed to be linear but sometimes are not for sugars (7). The exponent in this equation (the slope in a logarithmic plot) indicates how rapidly sensation increases relative to concentration. Unfortunately, there is no standard or typical value for sucrose; published values for the exponent for simple sucrose solutions range from 0.5 to 1.3 (1, 7, 20, 28, 29, 31, 42). Available evidence suggests that the exponent does not vary among natural sugars (28,29). Pleasantness of sugar solutions does not increase monotonically with concentration; solutions containing 3 to 10 percent sucrose (0.1 to 0.3 mol) are usually the most preferred (3, 11, 16, 19, 29, 31, 33, 37). In this case, the equation given above must be modified to include a quadratic component.

Psychometric techniques require fewer assumptions than psychophysical techniques about people's ability to measure sensations; all that is required is that a person say one sensation is greater than another. Indeed, subjects are usually assisted by a set of verbal descriptors for degrees of sensation (for example, very pleasant, slightly pleasant, neither pleasant nor unpleasant, slightly unpleasant, very unpleasant). Subjects may be forced to choose responses on a 5- to 9-point scale or mark a spot on a continuous line (30). The disadvantage of psychometric techniques is that they do not provide estimates of the numerical relationships between different tastes (for example, is one solution twice as sweet as another).

It is common for researchers to use both psychometric and psychophysical techniques, but systematic comparisons of their relative merits have not been conducted for sweet taste. Food technologists may find psychometric techniques to be the most useful because the goal is usually to find the sucrose concentration that produces the best-tasting food; numerical relationships between various levels of sucrose are often not needed. Psychometric techniques are also more suitable for large-scale surveys.

Additional information on measurement techniques is available in several places (see 2, 14, 25, 30).

Validity of Measures

Measurements of taste taken in the laboratory can be reliable and even valid for that context without having any relationship to the real world. A striking example is the finding of Mattes and Mela (27) that "preferred sweetener level in coffee corresponded to 5 teaspoons of sugar in a 6-oz serving."

An assumption underlying many psychophysical studies is that humans respond to absolute levels of sweetness. However, people tend to compare new sensory impressions to those they have just experienced. Thus, a given solution will be described as sweeter when it is tasted after dilute sucrose solutions than after tasting concentrated solutions, even when the subjects rinse their mouths with water between taste tests (37). Pleasantness judgments are affected in the same way as sweetness intensity (37). This factor should be taken into consideration when formulating new foods, especially if one knows anything about when the new food will be consumed. For example, a beverage that is expected to be consumed with a sweet snack should be tested along with sweet snacks.

The food technologist is primarily concerned with developing a food that people will consume often; measurements of pleasantness are useful only insofar as they predict intake. Unfortunately, little is known about how well pleasantness ratings correlate with intake

(25). One report suggests that taste-and-spit preference tests give a higher optimal sugar level in yogurt than do consumption tests (24). The only good data on the relationship between short- and long-term tests come from animals. Animal studies have shown that brief preference tests do not predict long-term consumption (23,46). The main reason short- and long-term test results disagree in animals almost certainly applies to humans; sugars affect bodily processes after they have been consumed. Such postingestive effects of sugars (caloric, osmotic, metabolic, and the like) take time to influence intake whereas taste is perceived rapidly.

Differences Among Sweeteners

No one distinguishes different kinds of sweetness. Indeed, the scientific literature suggests that psychophysical "laws" are identical for different carbohydrate sweeteners. Nevertheless, people can discriminate between different sweeteners (39). Distinguishing artificial sweeteners and natural sugars is usually easy because the former can have bitter or metallic tastes and produce less viscous solutions. It is probable that the most important factors differentiating sugars from artificial sweeteners in simple solutions are their rapidity of onset and duration of action (4), but detailed comparisons have not been published. This factor may be particularly important in foods having a low-moisture content because rate of solution and viscosity of the solution in saliva could substantially affect the taste of a sweet food. Once a sugar is incorporated into a food, chemical reactions can occur (e.g., Maillard reaction) that will vary according to the sugar used. No general rules can be given, but in any complex food system, substituting one sweetener for another will usually alter the flavor of the food even if sweetness is controlled.

Interactions Between Sweet Taste and Other Sensations

With respect to taste thresholds and intensity of sensation, different tastes tend to suppress one another (19, 21, 22). Thus, a mixture of bitter and sweet substances tastes less bitter and less sweet than the components would taste if present in separate solutions. One exception to this rule of mutual suppression is that 0.04 to 0.35 percent sodium chloride can enhance the sweetness of dilute sucrose solutions (21); this is probably due to the slight sweetness of sodium chloride. Whether pleasantness is influenced by mixture suppression has not been adequately examined, but mixing certain tastes might enhance rather than suppress pleasantness. Sweet-sour mixtures and sweet-bitter mixtures (soft drinks and fruit juices) are more frequently consumed than pure sweet solutions, even though sour and bitter tastes alone are rarely liked.

Mixtures of sweeteners generally show additive or synergistic effects; an example of the latter is the substantial improvement in the taste of saccharin resulting from the addition of small amounts of carbohydrate sweeteners (41). De Graaf and Frijters (8) have proposed two simple rules for predicting the sweetness of mixtures of carbohydrate sweeteners: (1) the sweetness of a mixture of two sugars is intermediate to the sweetness of the individual components at the same total molarity as the mixture, and (2) as the proportion of one sugar increases, the sweetness of the mixture approaches that of the more abundant sugar at the same total molarity of the mixture. These rules apply to sweetness intensity; pleasantness has not been evaluated in this way.

Less is known about how sucrose may affect odor-based flavors. Physiochemical effects may be the most important factors in this situation. The presence of substantial amounts of sugar in a solution will alter the solubility of volatile substances. The degree to which this occurs depends on so many different factors (specific chemicals, temperature, other substances in food, etc.) that it is not possible to make any firm generalizations about what effects will occur (21). However, it is probable that these effects will be greater for sucrose and other sugars than for artificial sweeteners because the latter substances are generally used in small amounts.

Interactions between sucrose and fat content have been studied in milk and cream (11, 12, 13). Magnitude estimates of sweetness were not influenced by the fat content of the milk, but preference ratings for a given level of sweetness were greater when the fat content was high (11, 12, 13). In other words, adding a given amount of sucrose to skim milk did not increase preference very much, whereas adding the same amount of sucrose to heavy cream substantially increased preference.

The taste of sugar may also be modified by purely physical factors such as temperature. Assessing the effects of temperature is complicated because mouth temperature tends to change when exposed to warm or cold solutions. This factor can be controlled by having people rinse their mouths with warm or cold solutions until mouth temperature stabilizes. When this is done, it is found that mouth temperature is a more important determinant of taste intensity than is solution temperature (17, 18). Cooling the tongue or the solution reduces the perceived intensity of sucrose, glucose, fructose, and aspartame (18). The taste of saccharin, however, is not strongly affected by either mouth or solution temperature (18). These factors could be put to use if one knows when a given food will be eaten; sweet solutions may taste very different when consumed following a hot meal rather than after a cool one.

Increasing solution viscosity usually decreases perceived sweetness and ability of people to detect sweetness (34). It is probable that this reflects reduced diffusion of sugar molecules in viscous solutions.

Differences Among People

People differ greatly in the degree to which they like the taste of sucrose and the level of sucrose that they consider to be most pleasant (6, 24, 27, 35). The significance and the causes of these individual differences are imperfectly understood. Some evidence indicates that people who like high levels of sugar in one food tend to like high levels of sugar in other foods (6, 35). However, individual differences in simple taste tests do not strongly correlate with habitual sugar usage (27, 44).

Several factors that might cause individual differences have been examined; these are listed in Table 6-1. A great deal of attention has been given to possible differences between lean and obese people (6, 12, 13, 43) on account of the notion that sugar intake is somehow related to obesity (10, 36). Consistent differences between lean and obese individuals have not been established.

The existence of large individual differences could be taken advantage of by marketing different lines of food products containing high, low, or intermediate amounts of sucrose.

Pleasantness of Sucrose

Whenever human beings have been able to do so, they have chosen to add an increased percentage of their calories in the form of sucrose and other sugars. Why do people do this? It is commonly assumed that people consume sucrose because it tastes good. However, this answer merely hides the question because it fails to address the issue of why people like sucrose. It is the contention of the author that a basic understanding of this issue is essential for the rational development of new foods.

Table 6-1. Reasons for Differences Among People in Liking for Sucrose

Factor	Reference
Long-Term Stable Factors	
Race, ethnicity	(3,19)
Gender	(6,44)
Age	(6,9,32)
Body weight or obesity	(6,12,13,43)
Transient Factors	
Intake of sweet foods	(15,38,45)
Hunger	(33,38,43)
Experience with sweet foods (learning)	(5)
State of health	(26,40)

In order to understand the problem, it is necessary to understand the concept of taste vs. post-ingestive effects. The postingestive effects of a sugar are those exerted by the sugar after it has been consumed. Postingestive effects include relief from hunger, altered stomach emptying, stimulation of gastrointestinal secretions, and repletion of glycogen stores. Taste affects one's immediate response to a food. Postingestive effects predominate late in a meal. If one repeatedly encounters a food with a certain postingestive effect, one will learn to associate the flavor of the food with that effect. For example, people who eat novel foods shortly before cancer chemotherapy, which induces feelings of malaise, subsequently dislike the test foods (26).

Some authors have suggested that there is a mathematical relationship between sweetness and pleasantness such that a given level of sweetness is associated with a definite level of pleasantness (16, 29), apparently regardless of postingestive effects. Thus, the perceived sweetness of a food should be more important than the sugar content per se. The range of situations in which this theory has been tested is too small to give much confidence in its validity, but it is an intriguing concept. An unstated assumption is that liking for sweetness is either innate or acquired early in life; in either case, liking of sweetness is thought to be a permanent trait having little relationship to experience. This would make life easy for food technologists because they could assume that if a person likes a candy bar today, that person will like the same candy bar equally well next year. Furthermore, different sweeteners could be used in the same product, provided that the taste of the product was the same.

Animal studies indicate that preference for foods is determined, in part, by the postingestive consequences of the foods (5, 23). Some studies with humans are consistent with this idea (5, 26), but the evidence is not as strong as it is with animals. The degree to which this applies to sugars is presently unknown. Most researchers have focused on mildly preferred or avoided flavors, and hence have avoided examining sweet tastes. Furthermore, we do not know which postingestive effects of sugars might be rewarding (fullness of stomach, repletion of liver glycogen, etc. [see 36]). If postingestive effects are indeed important, tests of new products will have to be conducted over lengthy periods in order to ensure that postingestive effects are allowed to influence liking and intake. Furthermore, altering the sweetener could alter postingestive effects and hence intake even when taste is unchanged.

At present it is impossible to say which view is closer to the truth. Nevertheless, the possibility that postingestive effects influence liking is sufficiently strong that precautions should be taken whenever new foods are developed. These precautions include testing over a prolonged period of time (days or months) and only using sweeteners that have a good track record for human acceptance (for instance, sucrose).

References

1. Bartoshuk, L.M. "Taste mixtures: is mixture suppression related to compression?" *Physiol. Behav.* 14:643–649. 1975.
2. Bartoshuk, L.M. and Marks, L.E. "Ratio Scaling." In H.L. Meiselman and R.S. Rivlin (eds.) *Clinical Measurement of Taste and Smell*, pp. 50–65. Macmillan, New York. 1986.
3. Bertino, M. and Chan, M. "Taste perception and diet in individuals with Chinese and European ethnic backgrounds." *Chemical Senses* 11:229–241. 1986.
4. Birch, G.G. and Latymer, Z. "Intensity/time relationships in sweetness: evidence for a queue hypothesis in taste chemoreception." *Chemical Senses* 5:63–78. 1980.
5. Booth, D.A. "Food-conditioned eating preferences and aversions with interoceptive elements: conditioned appetites and satieties." *Annals of the New York Academy of Sciences* 443:22–41. 1985.

6. Connor, M.T. and Booth, D.A. "Preferred sweetness of a lime drink and preference for sweet over non-sweet foods, related to sex and reported age and body weight." *Appetite* 10:25–35. 1988.
7. Curtis, D.W., Stevens, D.A. and Lawless, H.T. "Perceived intensity of the taste of sugar mixtures and acid mixtures." *Chemical Senses* 9:107–120. 1984.
8. De Graaf, C. and Frijters, J.E.R. "Sweetness intensity of a binary sugar mixture lies between intensities of its components, when each is tasted alone and at the same total molarity as the mixture." *Chemical Senses* 12:113–129. 1987.
9. Desor, J.A., Greene, L.S., and Maller, O. "Preferences for sweet and salty in 9-to-15-year-old and adult humans." *Science* 190:686–687. 1975.
10. Drewnowski, A. "Sweetness and obesity." In Dobbing, J. (ed.) *Sweetness*, pp. 177–192. Springer-Verlag, London. 1987.
11. Drewnowski, A. and Greenwood, M.R.C. "Cream and sugar: human preferences for high-fat foods." *Physiol. Behav.* 30:629–633. 1983.
12. Drewnowski, A., Grinker, J.A., and Hirsch, J. "Obesity and flavor perception: multidimensional scaling of soft drinks." *Appetite* 3:361–368. 1982.
13. Drewnowski, A., Brunzell, J.D., Sande, K., Iverius, P.H., and Greenwood, M.R.C. "Sweet tooth reconsidered: taste responsiveness in human obesity." *Physiol. Behav.* 35:617–622. 1985.
14. Engen, T. "Classical psychophysics: humans as sensors." In Meiselman, H.L. and Rivlin, R.S. (eds.) *Clinical Measurement of Taste and Smell*, pp. 39–49. Macmillan, New York. 1986.
15. Esses, V.M. and Herman, P. "Palatability of sucrose before and after glucose ingestion in dieters and nondieters." *Physiol. Behav.* 32:711–715. 1984.
16. Frijters, J.E.R. "Sensory sweetness perception, its pleasantness, and attitudes to sweet foods." In Dobbing, J. (ed.) *Sweetness*, pp. 67–80. Springer-Verlag, London. 1987.
17. Green, B.G. and Frankmann, S.P. "The effect of cooling the tongue on the perceived intensity of taste." *Chemical Senses* 12:609–619. 1987.
18. Green, B.G. and Frankmann, S.P. "The effect of cooling on the perception of carbohydrate and intensive sweeteners." *Physiol. Behav.* 43:515–519. 1988.
19. Greene, L.S., Desor, J.A., and Maller, O. "Heredity and experience: Their relative importance in the development of taste preference in man." *J. Comp. Physiol. Psychol.* 89:279–284. 1975.
20. Kocher, E.C. and Fisher, L. "Subjective intensity and taste preference." *Perceptual and Motor Skills* 28:735–740. 1969.
21. Lachmann, A. *The Role of Sucrose in Foods.* International Sugar Research Foundation, Bethesda, MD. 1975.
22. Lawless, H.T. "Evidence for neural inhibition in bittersweet taste mixtures." *J. Comp. Physiol. Psychol.* 93:538–547. 1979.

23. LeMagnen, J. "Palatability: concept, terminology and mechanisms." In Boakes, R.A., Popplewell, D.A., and Burton, M.J. (eds.) *Eating Habits*, pp. 131–154. John Wiley, Chichester. 1987.
24. Lucas, F. and Bellisle, F. "The measurement of food preferences in humans: do taste-and-spit tests predict consumption?" *Physiol. Behav.* 39:739–743. 1987.
25. Mattes, R.D. "Reliability of psychophysical measures of gustatory function." *Perception & Psychophysics*. 4:107–114. 1988.
26. Mattes, R.D., Arnold, C., and Boraas, M. "Management of learned food aversions in cancer patients receiving chemotherapy." *Cancer Treatment Reports* 71:1071–1078. 1987.
27. Mattes, R.D. and Mela, D.J. "Relationships between and among selected measures of sweet-taste preference and dietary intake." *Chemical Senses* 11:523–539. 1986.
28. Moskowitz, H.R. "Ratio scales of sugar sweetness." *Perception & Psychophysics* 7:315–320. 1970.
29. Moskowitz, H.R. "The sweetness and pleasantness of sugars." *Am. J. Psychol.* 84:387–405. 1971.
30. Moskowitz, H.R. "Overview of hedonic measurement." In Meiselman, H.L. and Rivlin, R.S. (eds.) *Clinical Measurement of Taste and Smell*, pp. 66–85. Macmillan, New York. 1986.
31. Moskowitz, H.R., Kluter, R.A., Westerling, J., and Jacobs, H.L. "Sugar sweetness and pleasantness: evidence for different psychological laws." *Science*. 184:583–585. 1974.
32. Murphy, C. "Taste and smell in the elderly." In Meiselman, H.L. and Rivlin, R.S. (eds.) *Clinical Measurement of Taste and Smell*, pp. 343–371. Macmillan, New York. 1986.
33. Pangborn, R.M. "Influence of hunger on sweetness preferences and taste thresholds." *Am. J. Clin. Nutr.* 7:280–287. 1959.
34. Pangborn, R.M. "Selected factors influencing sensory perception of sweetness." In Dobbing, J. (ed.) *Sweetness*, pp. 49–66. Springer-Verlag, London. 1987.
35. Pangborn, R.M. and Giovanni, M.A. "Dietary intake of sweet foods and of dairy fats and resultant gustatory responses to sugar in lemonade and to fat in milk." *Appetite* 5:317–327. 1984.
36. Ramirez, I. "When does sucrose increase appetite and adiposity?" *Appetite* 9:1–19. 1987.
37. Riskey, D.R., Parducci, A., and Beauchamp, G.K. "Effects of context in judgments of sweetness and pleasantness." *Perception & Psychophysics* 26:171–176. 1979.
38. Scherr, S. and King, K.R. "Sensory and metabolic feedback in the modulation of taste hedonics." *Physiol. Behav.* 29:827–832. 1982.
39. Schiffman, S.S., Reilly, D.A., and Clark, T.B., III. "Qualitative differences among sweeteners." *Physiol. Behav.* 23:1–9. 1979.

40. Settle, R.G. "Diabetes mellitus and the chemical senses." In Meiselman, H.L. and Rivlin, R.S. (eds.) *Clinical Measurement of Taste and Smell*, pp.487–513. Macmillan, New York. 1986.
41. Smith, J.C., Foster, D.F., and Bartoshuk, L.M. "The synergistic properties of pairs of sweeteners." In Barker, L.M. (ed.) *The Psychobiology of Human Food Selection*, pp. 123–138. AVI/Van Nostrand Reinhold, New York. 1982.
42. Stevens, S.S. "Sensory scales of taste intensity." *Perception & Psychophysics* 6:302–308. 1969.
43. Thompson, D.A., Moskowitz, H.R., and Campbell, R.G. "Effects of body weight and food intake on pleasantness ratings for a sweet stimulus." *J. Appl. Physiol.* 41:77–83. 1976.
44. Tuorila-Ollikainen, H. and Mahlmäki-Kulltanen, S. "The relationship of attitudes and experiences of Finnish youths to their hedonic responses to sweetness in soft drinks." *Appetite* 6:115–124. 1985.
45. Wooley, O.W., Wooley, S.C., and Dunham, R.B. "Calories and sweet taste: effects on sucrose preference in the obese and nonobese." *Physiol. Behav.* 9:765–768. 1972.
46. Young, P.T. "Palatability: the hedonic response to foodstuffs." In Code, C.F. (ed.) *Handbook of Physiology, Section 6. Alimentary Canal, Vol. 1. Control of Food and Water Intake*, pp. 353–366. American Physiological Society, Washington, DC. 1967.

7

Sugar in the Body

Theodore Cayle, Ph.D.*

For the purposes of this discussion, the term sugar refers to sucrose. Other biological sugars will be designated by their scientific name, such as glucose or fructose, or "sugars" will be used in the plural to indicate all carbohydrate sweeteners.

The formula for sugar is shown in Figure 7-1. It can be seen that sucrose is nonreducing since the aldehyde group of the D-glucosyl component and the ketone group of the D-fructosyl moiety are bonded chemically. The former is used to form the glycosidic linkage between glucose and fructose, while the latter is utilized in the furanosyl ring of fructose.

Sucrose has the empirical formula $C_{12}H_{22}O_{11}$ and a molecular weight of 342.30. Sucrose crystals are monoclinic prisms having a density of 1.588. A 26 percent solution has a density of 1.108175 grams per milliliter at 20°C. Sucrose is optically active with a specific rotation $[\alpha]_D^{20}$ of +66.53°. Its melting point is 185 to 186°C; when heated above 200°C, sucrose decomposes. The refractive index is 1.3740 for a 26 percent solution at 20°C. Sucrose is soluble in water and ethanol, saturated solutions at 20°C being 67.09 and 0.90 percent by weight, respectively. Sucrose is only slightly soluble in methanol and is insoluble in ether and chloroform (1).

Production in Green Plants

Sucrose is produced by an intricate sequence of biochemical reactions that occur during the distinct processes of photosynthesis and

*The late Theodore Cayle, Ph.D., was Vice President, Scientific Affairs, The Sugar Association, Inc., Washington, D.C.

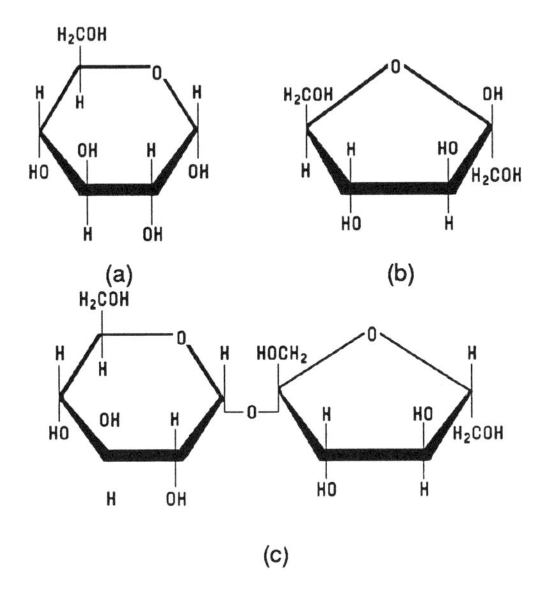

Figure 7-1. (a) α-D-glucopyranose; (b) β-D-fructofuranose; (c) sucrose (α-D-glucopyranosyl-β-D-fructofuranoside or β-D-fructofuranosyl-α-D-glucopyranoside).

sugar formation. During photosynthesis, carbon dioxide, in the presence of chlorophyll and sunlight, is added (fixed) to a plant-specific biochemical constituent. The particular molecule resulting from carbon dioxide fixation is enzymatically transformed into sucrose precursor and energy. Conversion of sugar precursor into sucrose is accomplished during sugar formation. Neither chlorophyll nor sunlight is required for sucrose formation. Thus, sugar formation occurs as readily in the dark as in sunlight.

Sugar Beet

Carbon dioxide is fixed to a molecule of ribulose-1,5-diphosphate (RuDP) during photosynthesis in the sugar beet. The resultant 6-carbon diphosphate intermediate is unstable and immediately, and irreversibly, breaks down to two molecules of 3-phosphoglyceric acid (3PG). Following enzymatic transformation (two reactions energized by ATP and NADPH) of 3PG to glyceraldehyde-3-phosphate (GAP), an equilibrium mixture of GAP and dihydroxyacetone phosphate (DHAP) is established. Since 3PG is a 3-carbon compound and is the first compound isolatable after carbon dioxide fixation, sugar beets are referred to as C_3 plants. The photosynthetic process

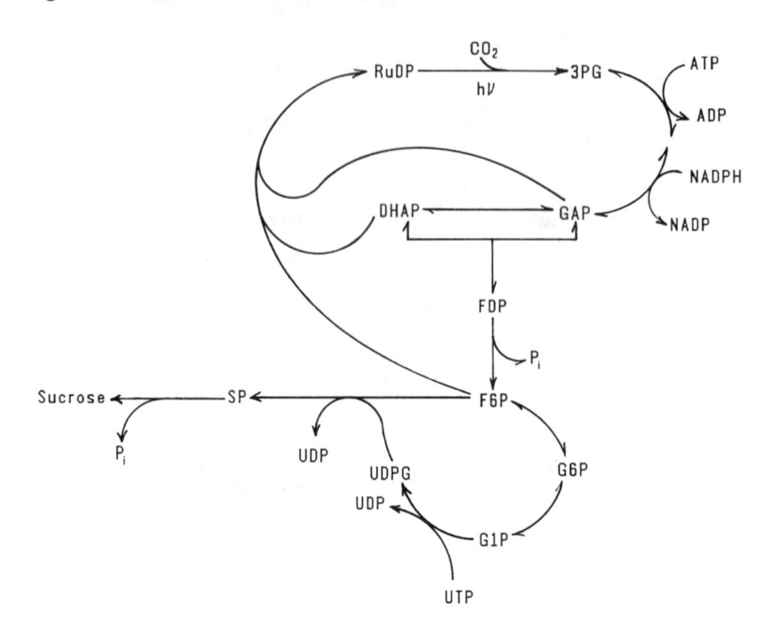

Figure 7-2. Summarized C_3 biosynthetic production of sucrose in sugar beets (abbreviations defined in chapter text; → signifies irreversible reaction; ⇄ signifies equilibrium mixture of materials).

of any green plant in which 3PG is the first product of carbon dioxide fixation is termed a C_3 pathway.

Sucrose formation begins with the condensation of GAP and DHAP to fructose-1,6-diphosphate (FDP) and subsequent irreversible hydrolysis of FDP to fructose-6-phosphate (F6P) and inorganic phosphate (P_i). Fructose-6-phosphate is the molecule from which sucrose is synthesized. A portion of the F6P is converted sequentially to glucose-6-phosphate (G6P), glucose-l-phosphate (G1P), and uridine diphosphate glucose (UDPG). The resultant UDPG is transferred to additional F6P to form sucrose phosphate (SP) and uridine diphosphate (UDP). Free sucrose is formed by a highly irreversible reaction, and is the first product of photosynthesis that is free of high-energy phosphate bonds.

The formation of sucrose by a C_3 biosynthetic pathway is summarized in Figure 7-2. It should be noted that F6P, GAP, and DHAP are the biochemical precursors for the multistep formation of RuDP. Figure 7-2 is an abbreviation at best. Specifics are obtainable from any detailed text on plant biochemistry or review article on photosynthesis.

Sugarcane

During photosynthesis in sugarcane and other C_4 plants, carbon dioxide is fixed initially to a molecule of phosphoenolpyruvate (PEP) to form the 4-carbon molecule, oxaloacetate (1,2). The oxaloacetate is converted to malate and transferred to bundle sheath cells where sugar formation occurs (2). Upon entering the bundle sheath cell, malate is enzymatically transformed to pyruvate and carbon dioxide. Pyruvate is transported back to mesophyll cells and, after conversion into PEP, is available for the next cycle of photosynthetic CO_2 fixation. The carbon dioxide released from malate in the sheath cell combines with RuDP to eventually form sucrose by the same mechanism shown in Figure 7-2. C_4 plants such as sugarcane synthesize carbohydrate at a faster rate, thus functioning more efficiently at the high light intensities of their environment than C_3 plants such as sugar beets (2).

Utilization of Sugar in Humans

According to Cahill et al. (3), the central nervous system requires approximately 140 grams of glucose per day, with an additional 40 grams needed by red blood cells and other tissues. Therefore, sugar in the human diet contributes to energy needs, if the mechanism for readily converting the sucrose molecule into glucose and fructose is present. At one time it was thought that sugar was hydrolyzed within the intestinal lumen. However, Crane and his colleagues (4–6) established that sucrase activity was located in the brush border membrane of the intestine.

Whereas human intestinal carbohydrases develop at various stages of fetal growth, Dahlqvist et al. (7) have shown that the enzyme responsible for sucrose hydrolysis is one of the first to appear, being fully active within three to ten weeks of gestation. Since entry into the interior of the intestinal cell is reserved for monosaccharides, it is essential that sucrase be functional. Its absence or malfunction not only prevents the cells from using the required glucose and fructose but can lead to osmotic irregularities, bacterial fermentations, and increased peristalsis, resulting in diarrhea, malabsorption, and flatus, from the nonhydrolyzable sucrose remaining within the intestinal lumen.

Glucose requires active transport to cross cell membranes, while fructose crosses cell membranes by carrier-mediated diffusion. The rate of monosaccharide absorption by diffusion mechanisms is significantly lower than the rate of absorption by active transport (8).

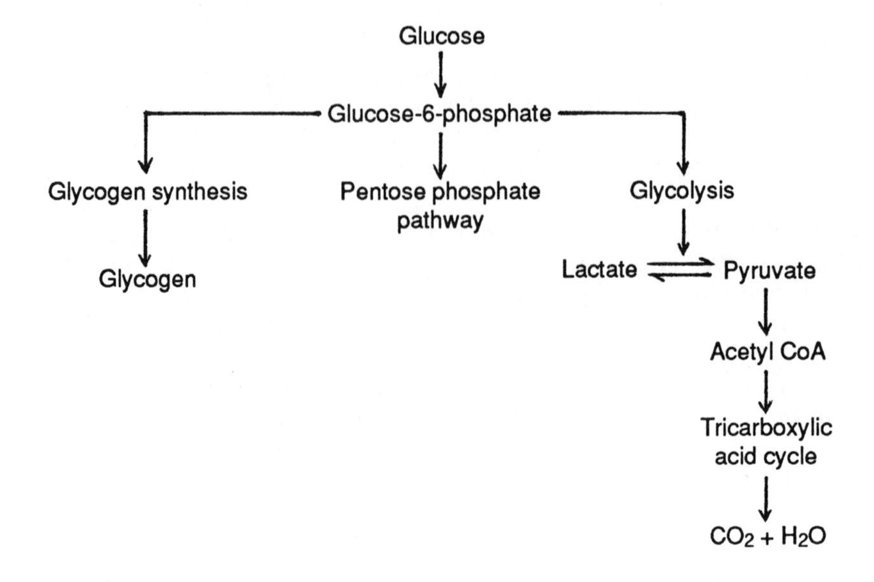

Figure 7-3. The metabolic fate of glucose (10). Reprinted by permission of Springer-Verlag, Heidelberg.

The exact molecular mechanism is unknown, but since glucose must cross against a concentration gradient, energy is required. Sodium ions appear to be required for the active transport of glucose. The sodium-dependent glucose carrier is able to achieve transport against a gradient by employing energy in the form of ATP and coupling to the transcellular flux of sodium (9).

The metabolic fate of glucose is represented in Figure 7-3. Glucose metabolism begins after phosphorylation to glucose-6-phosphate. The utilization of glucose-6-phosphate depends on cell type and its immediate biochemical requirements (10). Glycolysis, with subsequent oxidation of pyruvate to carbon dioxide, water, and energy, is the major mechanism by which glucose is metabolized. When glucose is not required for immediate energy demands, the extra glucose is converted preferentially to glycogen, the energy reserve of human metabolism. After glycogen requirements have been satisfied, the remaining glucose is metabolized by the pentose phosphate pathway. The pentose phosphate pathway is responsible for as much as 30 percent of the glucose metabolized in the liver, and more than 30 percent in fat cells (10). After one cycle of this pathway, twelve hydrogen ions are produced per mole of glucose. Depending

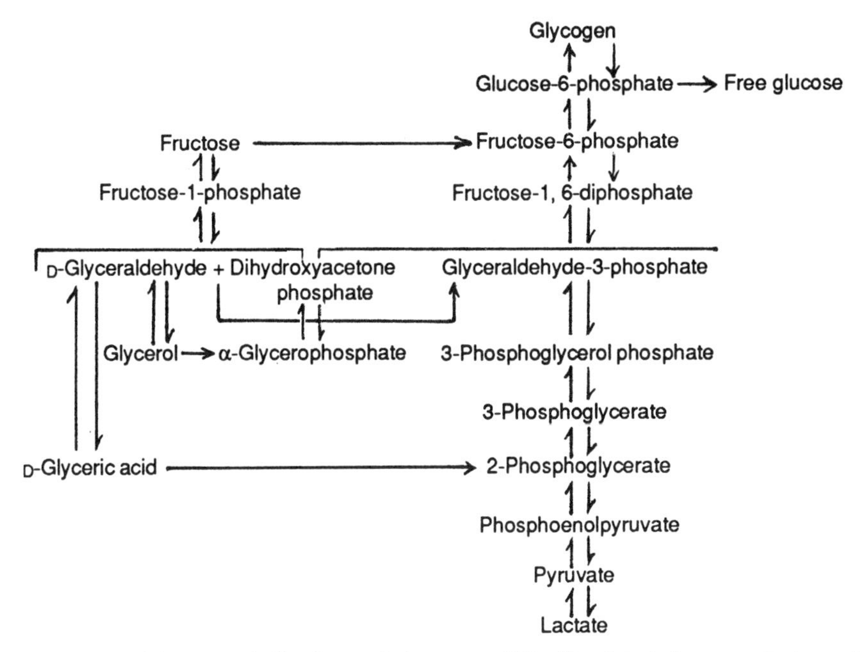

Figure 7-4. The metabolic fate of fructose (10). Reprinted by permission of Springer-Verlag, Heidelberg.

on the energy state of the cell, the resultant hydrogen ions are utilized for energy production or fatty acid metabolism. Approximately 40 percent of the total potential energy is generated during human metabolism of one mole of glucose (11).

The liver, kidney, and small intestine are the major sites of human fructose metabolism, with the liver predominating. Fructose is absorbed by carrier-mediated diffusion (10) at a rate between those of active transport (glucose) and passive diffusion (mannose). Up to 90 percent of ingested fructose is absorbed as fructose in humans. The metabolic fate of absorbed fructose is outlined in Figure 7-4.

Fructose is converted to glucose, lactate plus pyruvate, glycogen, and triglycerides following absorption (10). Conversion to glucose is the major metabolic fate, where conversions up to 70 percent in humans have been reported (12,13). The majority of the absorbed fructose is converted to fructose-1-phosphate. Fructose-6-phosphate is a minor metabolite in the liver because the specificity and activity of endogenous enzymes favor glucose over fructose (10). As shown in Figure 7-4, fructose-1-phosphate is transformed into an equilibrium mixture of glyceraldehyde, dihydroxyacetone, and glyceraldehyde-3-phosphate. Glyceraldehyde-3-phosphate is the molecule by which fructose is transformed to glucose-6-phosphate for conversion to glucose and, to a lesser extent, glycogen. The resultant glucose is metabolized by one of the three alternative pathways shown in Figure 7-3. When immediate energy needs are satisfied, glyceraldehyde-3-phosphate is enzymatically converted to lactate for future metabolic demands (10). Fructose may also be converted to glycerol and glyceric acid (Figure 7-4). If all ingested fructose were transformed to glucose or lactate (pyruvate), the potential energy generated during oxidative fructose metabolism would equal that derived from the complete oxidation of glucose. Thus, an energy yield of 40 percent would be expected from the complete oxidation of sucrose in the biochemically normal human.

The postprandial concentration of glucose circulating in the blood of a normal individual rises from its fasting value of about 80 milligrams per deciliter to 140 milligrams per deciliter in the first hour, and then returns to the fasting value within two to three hours. Regulation of the blood glucose concentration is mediated primarily by two hormones, insulin and glucagon. These provide the control experienced by all normal individuals, resulting in the maintenance of blood glucose levels within limits that prevent either hyperglycemia (high blood glucose) or hypoglycemia (low blood glucose). Individuals with diabetes mellitus can have hyperglycemia after ingesting sugar due to a lack of functioning insulin or insulin receptors. Such individuals must take insulin routinely in order to metabolize glucose and prevent its excretion in the urine.

Sugar and Health

The FASEB Report (1976)

For regulatory purposes (14), most foods and food ingredients are placed in one of two categories: safe or not safe. For those substances used prior to January 1, 1958, the Food, Drug and Cosmetic Act [21 USC 321(s)] provides for declaring these as generally recognized as safe (GRAS). Those substances not considered GRAS must be the subject of a Food Additive Petition, which includes extensive testing procedures, before they are approved for use in foods. The term GRAS, according to the Code of Federal Regulations 21 CFR 121.1 as revised April 1, 1976, means general recognition of safety by experts qualified by scientific training and experience to evaluate the safety of substances on the basis of scientific data derived from published literature.

As part of a general review of GRAS substances, the Food and Drug Administration (FDA), part of the U.S. Department of Health and Human Services, contracted with the Life Sciences Research Office of the Federation of American Societies for Experimental Biology (FASEB) to review the GRAS status of sucrose. FDA provided two scientific literature reviews, prepared in 1973 and 1974, that summarized the world's scientific literature from 1920 through 1974. An exhaustive search of the available literature was made by this Select Committee on GRAS Substances, public hearings were held, and the resulting deliberations were published in 1976 under the title "Evaluation of the Health Aspects of Sucrose as a Food Ingredient" (14).

The report points out that standards of identity and purity for sugar are listed in the U. S. Pharmacopeia. Jukes (15) has called attention to the fact that "this emphasizes the purity of sugar as sold for food use. By a perversion of logic, this very purity is raised as an objection to sugar by the natural food industry."

In addition to acute and chronic studies, the following special studies were evaluated:

a. Carcinogenicity
b. Teratogenicity
c. Atherosclerosis
d. Diabetes
e. Dental caries

The conclusions of the Select Committee were

1. Reasonable evidence exists that sucrose is a contributor to the formation of dental caries when used at the levels that are now current and in the manner now practiced.

2. Other than the contribution made to dental caries, there is no clear evidence in the available information on sucrose that demonstrates a hazard to the public when used at the levels that are now current and in the manner now practiced. However, it is not possible to determine, without additional data, whether an increase in sugar consumption—that would result if there were a significant increase in the total of sucrose, corn sugar, corn syrup, and invert sugar, added to foods—would constitute a dietary hazard.

The term "sugar" used by the Select Committee in this instance is not restricted to sucrose.

The FASEB report (SCOGS-69), as well as an additional report on the evaluation of corn sugar, corn syrup, and invert sugar (SCOGS-50), provided FDA with the assurances regarding the safety of sucrose and the other sweeteners to enable it to attempt to affirm the GRAS status of these food ingredients. Accordingly, FDA published a GRAS affirmation proposal in the *Federal Register* in November, 1982 (47 CFR 53917 and 53923).

FDA Task Force on Sugars (1986)

Since a number of comments were received by FDA, indicating that significant new data on the impact of sugar and other sweeteners on health had become available since the publication of the FASEB report in 1976, Dr. Sanford A. Miller, Director of the FDA Center for Food Safety and Applied Nutrition, requested that a Sugars Task Force be established to review and interpret the recent literature applicable to the health effects of these sweeteners in food. The literature review encompassed dental caries, glucose tolerance, diabetes mellitus, lipids, cardiovascular disease including hypertension and atherosclerosis, behavior, and obesity. In addition, the Task Force determined consumption patterns derived from USDA disappearance data.

The Task Force summarized the most important studies in a particular area and determined whether the scientific data raised a significant question of safety relative to the normal human population.

Dental Caries

The etiology of dental caries is well known and extremely complex (16). There is a multiplicity of factors involved, including the microorganisms of the mouth that are capable of fermenting carbohydrates to lactic acid and of producing sticky polysaccharides, which form the foundation of plaque. Tooth structure also plays a role in providing a focal point for plaque attachment. The intensity of oral hygiene practices is still another component of the matrix influencing the potential for cariogenicity. Dietary practices, such as the frequency of snacking on sweet foods and the texture of these foods, also contribute to the potential for the production of dental caries.

Because of the above, the oft-quoted statement that sugar (sucrose) is responsible for dental caries is not only simplistic but misleading. All fermentable carbohydrates can contribute to the formation of dental caries if all other parameters of the equation are present. The proof of this is that although the consumption of sugar and other nutritive sweeteners has remained constant over the years in the U.S., dental caries have been dramatically reduced. An announcement by the National Institutes of Dental Research (NIDR) of the National Institutes of Health on June 21, 1988, called attention to the fact that half the school children in the United States have no tooth decay. According to the NIDR, American children now have 36 percent fewer dental caries than they did at the beginning of the 1980s. This was attributed to the widespread use of fluoride in community water supplies, toothpastes, and other forms. Thus, the modification of just one of the contributing parameters, the addition of fluoride to the smooth surfaces of teeth, has resulted in this remarkable achievement.

While the FDA report did identify sucrose as one of the fermentable carbohydrates that can contribute to dental caries, it is also obvious that this is but one risk factor. With proper oral hygiene and the inclusion of fluoride in drinking water and dentifrices, the impact of sucrose on dental caries can be reduced to insignificance.

Glucose Tolerance

Plasma glucose levels after oral loading are dependent upon the carbohydrate ingested, the physical state of the foodstuff, the form in which it is eaten, and the physical status of the individual.

It has been recognized for a number of years that different carbohydrate foods elicit different plasma glucose levels, or glycemic reponse. Glycemic reponse has been used to develop appropriate diets for the diabetic (17–20). Jenkins et al. (21) developed the concept of redefining glycemic response as a glycemic index of carbohydrates. The glycemic index of sucrose was defined as 90, relative to the standard of 100 for white bread. Jenkins et al. (22) also called attention to three qualifications required in the application of the glycemic index: 1) large individual variation in glycemic responses; 2) lack of agreement among different centers; and 3) lack of difference between mixed meals. Although Jenkins et al. did attempt to respond to this criticism, a number of investigators are still questioning the clinical benefit to be gained by designing meals based on the glycemic index (23–25).

A number of investigators have pointed out the importance of defining the physical state of the foodstuff when evaluating the glycemic response of different substances. A good example of the need to standardize all aspects of such studies was reported by Holm et al. (26). The degree of gelatinization of the starch fed to rats influenced plasma glucose and insulin responses.

The FDA Sugars Task Force evaluated a number of animal and human studies on the effects of sucrose and some monosaccharides on plasma glucose and insulin levels after oral loads. One long-range study in humans of particular interest was reported by Huttunen et al. (27). Two-year total replacement of sucrose by fructose or xylitol in the diets of volunteers did not affect glucose tolerance. Serum triglycerides, fasting blood glucose, and insulin levels were not influenced by this replacement.

The FDA Sugars Task Force also called attention to the work of Reiser's group at the U. S. Department of Agriculture on the effects of diet on glucose tolerance in individuals that were preselected for their exaggerated insulin reaction to oral sucrose loads. According to Reiser (28,29), these individuals are classified as "carbohydrate-sensitive" and are presumed to be genetically predisposed to abnormal reactions elicited by sugar ingestion.

The conclusion of the Task Force regarding Reiser's work was:

> These are complicated experiments and difficult to relate to current levels of sugars consumption. It does appear that a hypertriglyceridemic male population can be subselected on the basis of an exaggerated insulin response to sucrose loading and that this population may be adversely impacted by sucrose consumption if they consume their meals in a gorging pattern. The prevalence of such a population is unknown. Furthermore, there is no reason to believe that the effect observed would be restricted to sucrose *per se* and would not also be produced by any diet with a high glycemic index. Alternatively, this population may be particularly sensitive to fructose by virtue of an action on insulin metabolism.... This possibility cannot be ruled out at this time.

Another conclusion of the Task Force was:

> Reports on the effects of high-sucrose diets on glucose tolerance in humans have been contradictory, ranging from improved glucose tolerance, to no effect, to impaired glucose tolerance. In most of the recent studies, glucose

tolerance seems to be improved after the consumption of sucrose-containing diets. There is no evidence that high sugars consumption alone would increase the risk of developing decreased glucose tolerance and eventually diabetes in healthy individuals.

Diabetes Mellitus

Diabetes mellitus is a heterogeneous group of clinical disorders characterized by hyperglycemia, many times associated with lipid and protein metabolic pathogenesis. The disease results from the destruction of the beta cells of the islets of Langerhans in the pancreas, where insulin is produced, a defect at the insulin receptor or postreceptor level, or a defect in hepatic uptake of glucose (30).

Diabetes mellitus has been classified into several categories. The two most important categories are insulin-dependent diabetes mellitus (IDDM), or Type I diabetes, and noninsulin dependent diabetes mellitus (NIDDM), or Type II diabetes (31).

While IDDM typically manifests itself early in life and historically was termed "juvenile diabetes," it can appear at any age (30). Those with this form of the disease require at least one injection of insulin daily.

NIDDM most often has its onset in adulthood and has been termed "maturity-onset diabetes." The majority of the individuals with NIDDM are obese, and this form of the disease appears to have a strong genetic component. These individuals, in contrast to those with IDDM, have measurable or even excessive amounts of insulin in their plasma (31).

Diet control and exercise are the recommended therapy for individuals with NIDDM. Arky (30) takes issue with those who recommend the restriction of total carbohydrate for all diabetics. He points out that food containing complex carbohydrates do not produce glycemic responses in proportion to their mono- and disaccharide constituents and that there is no documented evidence that total carbohydrate intake should be disproportionately restricted in individuals with NIDDM. In fact, he refers to studies in which high-carbohydrate diets (55 to 80 percent of total calories as carbohydrate) improved glucose tolerance in NIDDM patients. Actually, his recommendation for carbohydrates as a percent of caloric intake for the NIDDM patient is the same as that recommended for a healthy individual, that is, between 50 and 60 percent as carbohydrate, with 15 percent or less as mono- and disaccharides.

The FDA Sugars Task Force report also cites studies in humans that show conflicting results. In some, a high-sucrose diet is reported to cause adverse effects, while in others an improvement in glucose tolerance was observed.

The Task Force concluded:

> There is no unequivocal evidence that high-sucrose consumption causes the onset of diabetes, and epidemiological studies were unable to show a connection between high sugars consumption and diabetes.... There is no evidence to support a change in the 1976 Select Committee on GRAS Substances report conclusion that the consumption of sugars is not related to diabetes other than as a nonspecific source of calories.

Lipids

The impact of diet on circulatory blood lipids has received considerable attention in recent years. Total serum cholesterol has been identified as a major component of the group of parameters associated with cardiovascular disease. It is now recognized that the type of lipoprotein transporting cholesterol in plasma is critical in determining if deposition in blood vessels is likely to occur. The three major carriers are the very low-density lipoproteins (VLDL), low-density lipoproteins (LDL), and high-density lipoproteins (HDL). The triglycerides content is highest in VLDL, intermediate in LDL, and lowest in HDL. Most plasma cholesterol is transported by the LDL, and it is this lipoprotein class that is mainly associated with cardiovascular disease (15). Claims have been made that sugar is responsible for elevated triglycerides and cholesterol. Ancel Keys addressed this issue in 1971 (32), identifying a number of studies that demonstrated little effect on the serum cholesterol level when subjects were fed diets extremely high in sucrose. Reiser (29) used a hypertriglyceridemic subset of his "carbohydrate-sensitive" patients to demonstrate elevated blood lipids in response to sucrose feeding. Recently, Reiser's group announced that it is the fructose component of sucrose that is responsible for the observed effects (33). However, the transfer of this information to the normal, healthy individual does not appear warranted. In the healthy person, any increase in blood triglycerides as a result of a high-sucrose diet is a transient effect in people of normal weight (34). Truswell (35) has also demonstrated that sucrose does not affect serum cholesterol levels under ordinary conditions.

The FDA's Sugars Task Force proposed a number of reasons why contradictory reports on blood lipids as a function of sugar intake continue to be published and why these discrepancies exist: 1) unknown role of genetic predisposition; 2) variable levels of physical activity; 3) interactive effects of other dietary components; 4) subject adaptation to changes in diet; 5) excessive amounts of sugar (up to 70 percent of caloric intake) in some experimental diets; 6) compromise from small experimental groups, short study duration, uncontrolled or nonisocaloric diets; 7) gorging patterns of meal distribution in some studies. The conclusion of the Task Force was, "Current levels of sugars consumption have not been demonstrated to be an adverse risk factor in terms of blood lipid and lipoprotein profiles for normal individuals."

Cardiovascular Disease

Yudkin (36,37) received considerable publicity with his theory that excess sugar consumption was responsible for cardiovascular disease (CVD). He developed this in a series of publications summarizing the evolutionary, historical, and epidemiological aspects of carbohydrate consumption. His indictment did not go unchallenged. Walker (38), Keys (32), Truswell (39), and Grande (40,41) repudiated Yudkin's claims, and the Puerto Rican study by Garcia-Palmieri et al. (42) also found no relationship between sugar intake and the incidence of CVD.

As concluded by the FDA Sugars Task Force, "There is no conclusive evidence that dietary sugars are an independent risk factor for coronary artery disease in the general population."

Behavior and Neurotransmitter Levels

Numerous scientific conferences have been organized to discuss the topic of diet and behavior. In general, there is some evidence to support a subtle association between behavior and certain dietary constituents. One of the topics generally discussed at these conferences is the generally held belief, based primarily on anecdotal reports, that ingestion of refined sugars may inordinately influence behavior, especially hyperactivity in children. Although experimental evidence supports the effect of diet on brain neurotransmitters, current data provide little support for anecdotal beliefs. A number of controlled studies demonstrating no link between hyperactivity and the ingestion of sugar have been published (43–49).

A physiological explanation for the role of carbohydrates in behavior was proposed by Wurtman and co-workers (50,51), after they observed a relationship between the ingestion of a protein or carbohydrate meal and brain serotonin level in rats. Serotonin (5-hydroxytryptamine) is a brain neurotransmitter that is derived from its amino acid precursor, tryptophan. Synthesis of serotonin is dependent upon the availability of tryptophan for transport across the blood-brain barrier. A key finding was that the availability of this amino acid, and thus brain serotonin, is increased following carbohydrate consumption. The effect of dietary sugars is mediated by the stimulatory action of insulin in clearing competing amino acids from the plasma and thereby allowing tryptophan relatively uninhibited access to the blood-brain barrier (52). Lyons and Truswell (53) reported that carbohydrates, particularly sucrose, are responsible for increased plasma tryptophan in healthy adult humans.

Since serotonin is a neurotransmitter that promotes sleep and lowers sensitivity to pain, one would expect sugar to have a calming, rather than stimulating, effect. In fact, such a response has been reported by a number of investigators (43,47) for children labeled by parents as hyperactive because of sugar ingestion.

The FDA Sugars Task Force concluded that "there is currently no scientifically validated evidence that indicates current levels of sugars consumption adversely affects behavior."

Obesity

Obesity is a major health problem because it has been identified as a risk factor for cardiovascular disease, hypertension, adult-onset diabetes, and certain forms of cancer. The excessive consumption of food in general, especially when accompanied by inadequate exercise, leads to the production and storage of body fat. It is the imbalance between calories consumed and calories expended that makes a normal person gain weight. Sugar and other carbohydrates yield approximately 4 calories per gram, protein 4 calories per gram, fats 9 calories per gram, and alcohol 7 calories per gram.

There appear to be obese subgroups. One subgroup has been shown to have a risk factor for body-weight gain because of a genetically inherited reduced rate of energy utilization (54). Another subgroup has been identified as "carbohydrate cravers," indulging in excessive carbohydrate intake in the form of snacks (55). Wurtman et al. (56) postulates that this craving is due to the need to produce serotonin, thereby bringing about a calming and relaxed feeling in the subjects. Oral administration of fenfluramine, a compound that selectively enhances serotonin-mediated neurotransmission, was shown (57) to cause a decrease in consumption of carbohydrate-rich snacks.

Of considerable significance to those concerned with the control of body weight as a function of diet have been the evidence accumulated over the last decade regarding the metabolic pathways of carbohydrate and fat, and the tracing of the path of dietary fat in the body. Danforth (58) summarized this work as follows:

1. Dietary fat can be entered into fat-stores of the body efficiently, with the expenditure of only 3 percent of the ingested calories, whereas it takes 23 percent of the ingested calories to store carbohydrate as fat.
2. The fatty-acid composition of human adipose tissue triglyceride closely reflects the pattern of fatty acids in the diet.
3. Fat stored in the body is derived mainly from fat, whereas dietary carbohydrate appears primarily as glycogen.

Earlier, Sims et al. (59) had found that lean subjects had great difficulty in gaining weight when overfed on a high-carbohydrate diet, compared with a similar group of lean subjects that gained weight with relative ease when the extra calories were supplied as fat.

Bjorntorp and Sjostrom (60) have suggested that less than 1 percent of carbohydrate in a mixed meal is incorporated into adipose tissue.

In the April-June 1988 edition of the U. S. Department of Agriculture Food and Nutrition Briefs, Conway reported that in a study with twenty-eight women on low- and high-fat diets, the high-fat diet contributed to a buildup of body fat.

Thus, with regard to obesity, it is becoming apparent that dietary fat, not dietary carbohydrate, should be the primary target of the weight watchers. The admonition by the Surgeon General to reduce the consumption of fats should be heeded by the obese in particular.

In summarizing their discussion of obesity, the Task Force concluded:

> Thus, the available data support the view that sugars do not have a unique role in the etiology of obesity. On the contrary, sugars and carbohydrates in general tend to be preferable as a source of energy expenditure in: 1) stabilizing resting metabolic rate during weight reduction; 2) increasing the size of BAT [brown adipose tissue] and the potential for food-induced thermogenesis; and 3) requiring more energy for fat deposition.

The Task Force report was published in 1986 (61). Nine hundred twenty-two publications were reviewed and evaluated. The overall assessment of the safety of sucrose and the other sweeteners was summarized by the Task Force as follows:

> The average daily intake for added sugars as a percentage of the daily calorie intake for the total population (11%) approximates the amount (10%) recommended by the Select Committee on Nutrition and Human Needs in its second edition of Dietary Goals for the United States.
>
> Evidence exists that sugars as they are consumed in the average American diet contribute to the development of dental caries.
>
> Other than the contribution to dental caries, there is no conclusive evidence on sugars that demonstrates a hazard to the general public when sugars are consumed at the levels that are now current and in the manner now practiced.

Surgeon General's Report on Nutrition and Health

The 700-page 1988 report (62), four years in the making, concentrates on the need for reducing fat consumption and recommends the eating of more complex carbohydrates and fiber. The two areas relating to sugar specifically mentioned are

1. Weight Control—"To reduce energy intake, limit consumption of foods relatively high in calories, fats, and sugars, and minimize alcohol consumption."
2. Tooth Decay—"Those who are particularly vulnerable to dental caries (cavities), especially children, should limit their consumption and frequency of use of foods high in sugars."

Current GRAS Status of Corn Sugar, Corn Syrup, Invert Sugar, and Sucrose

On November 7, 1988, the Food and Drug Administration published in the *Federal Register* (63) the reaffirmation of sugar as generally recognized as safe as a direct food ingredient. In this document, FDA also responded to the critics of its Sugars Task Force Report, once again reiterating the general safety of sugar:

On the issue of dental caries, both the Task Force report and the Select Committee on GRAS Substances reports concluded that the current level of consumption of sweeteners, and of the sugars they contain, contributes to the incidence of dental caries in the general population, but that this consumption is not the only factor contributing to the incidence of dental caries.

The Task Force also found that dental caries incidence in the United States has declined significantly since the Select Committee issued its report in spite of the fact that sugars consumption has remained unchanged over that period. The Task Force attributed this decline, in part, to preventative dental methods.

Both the Select Committee and the Task Force have concluded that there is no conclusive evidence that sugar consumption at present levels poses a health hazard to the general public, other than a contribution to dental caries.

Miscellaneous Publications

1. American Society for Clinical Nutrition, Inc., Task Force Report (1979) (64).

A committee of the American Society for Clinical Nutrition (ASCN) Task Force on the Evidence Relating Six Dietary Factors to the Nation's Health reviewed the role of carbohydrates, including sucrose, on human disease. The health issues addressed were atherosclerotic heart disease, diabetes mellitus, obesity, and dental disease.

The ASCN group came to the same conclusions as the FASEB Select Committee: that, other than its contribution to dental caries, sugar demonstrates no hazard to public health.

2. Report of the British Nutrition Foundation's Task Force: Sugars and Syrups, London: BNF, 1987 (65).

This document summarized the BNF Task Force deliberations. The Task Force came to the same conclusions as those found by the FDA Task Force report. Each report stated that sugar's contribution to dental caries is the sole health-related issue of any concern.

3. Position of the American Dietetic Association (66).

The American Dietetic Association's position, published in 1987, is, "Except in cases of inborn errors of metabolism, there is no evidence to indicate that sugar intake at current levels is a risk factor in any particular disease, other than dental caries."

4. American Council on Science and Health (1986) (34).

The American Council on Science and Health (ACSH) independently evaluated the scientific evidence on the health effects of sugar and published its report in 1986. The conclusion was,

> Based on its analysis of the scientific evidence, the American Council on Science and Health concludes that sugars do not pose a threat to health when consumed in the amounts that have been customary in the United States for the past fifty years, with the exception of the role that sugars and other carbohydrates play in promoting tooth decay.

References

1. Mead, G. P. and Chen, J. C. P. *Cane Sugar Handbook*, 10th ed. John Wiley and Sons, New York, NY, p. 20 (1977).
2. Clark, M. G. *Proceedings of World Sugar Research Organization Scientific Conference*, Australia, August 10–15, pp. 36–42 (1986).
3. Cahill, G. F., Jr., Owen, O. E., and Felig, P. *The Physiologist* 11: 97–102 (1968).
4. Miller, D. and Crane, R. K. *Biochim. Biophys. Acta* 52: 281–293 (1961).
5. Eichholz, A. and Crane R. K. *J. Cell Biol.* 26: 687–691 (1965).
6. Maestracci, D., Schmitz, J., Preiser, H. and Crane, R. K. *Biochim. Biophys. Acta* 323: 113–124 (1973).
7. Dahlqvist, A. and Lindberg, T. *Clin. Sci.* 30: 517 (1966).
8. Herman, R. H. in *Sugars in Nutrition*, ed. by Sipple, H. L. and McNutt, K. W. Academic Press, New York, NY, p. 152 (1974).
9. Crane, R. K. in *Physiological Effects of Food Carbohydrates*, ed. by Jeanes, A. and Hodge, J., American Chemcial Society, Washington, DC, pp. 2–19 (1975).
10. *Sucrose Nutritional and Saftey Aspects*, ed. by Vettorazzi, G., and Macdonald, I. Springer-Verlag, Heidelberg, pp. 35–48 (1988).
11. Pike, R. L. and Brown, M. L. In *Nutrition: An Integrated Approach*, 2nd ed. John Wiley and Sons, New York, pp. 490–531 (1975).
12. Ockerman, P. A. and Lundborg, H. *Biochim. Biophys. Acta* 105: 34–42 (1956).
13. White, W. and Landau, B. L. *J. Clin. Invest.* 44: 1200–1213 (1965).
14. FASEB. *Evaluation of the Health Aspects of Sucrose as a Food Ingredient.* National Technical Information Service, Springfield, VA (1976).
15. Jukes, T. H. *World Rev. Nutr. Diet* 48: 137–194 (1986).
16. Shaw, J. H., *N. Engl. J. Med.* 317: 996–1004 (1987).
17. Wagner, R. and Warkany, J. *Z. Kinderkeilkd* 44: 322 (1927).
18. Conn, J. W. and Newburg, L. H. *J. Clin. Invest.* 15: 665–671 (1936).

19. Otto, H., Bleyer, G., Pennartz, M., Sabin, G., Schauberger, G., and Spaethe, R. in *Symposium on Diatetik bei Diabetes Mellitus*, ed. by Otto, H. and Spaethe, R. Huber, Bern, Switzerland, pp. 41–50 (1973).
20. Otto, H. and Niklas, L. *Med. Hyg.* 38: 3424–3429 (1980).
21. Jenkins, D.J.A., Wolever, T.M.S., Taylor, R.H., Barker, H., Fielden, H., Baldwin, J.M., Bowling, A.C., Mewman, H.C., Jenkins, A.L., and Goff, D.V. *Am. J. Clin. Nutr.* 34: 362–366 (1981).
22. Jenkins, D.J.A., Wolever, T.M.S., and Jenkins, L.J. *Diabetes Care* 11: 149–159 (1988).
23. Coulston, A.M., Hollenbeck, C.B., and Reaven, G.M., *Am. J. Clin. Nutr.* 39: 163–165 (1984).
24. Hollenbeck, C.B., Coulston, A.M., and Reaven, G.M. *Diabetes Care* 9: 641–647 (1986).
25. Hollenbeck, C.B., Coulston, A.M., and Reaven, G.M. *Diabetes Care* 11: 323–329 (1988).
26. Holm, J., Lundquist, I., Bjorck, I., Eliasson, A.C., and Asp, N.G. *Am. J. Clin. Nutr.* 47: 1010–1016 (1988).
27. Huttunen, J.K., Makinen, K.K., and Scheinen, A. *Acta Odont. Scand.* 33: 239–245 (1976).
28. Reiser, S., Bohn, E., Hallfrisch, J., Michaelis, O.E. IV, Keeney, M., and Prather, E.S. *Am. J. Clin. Nutr.* 34: 2348–2358 (1981).
29. Reiser, S., Bickard, M.C., Hallfrisch, J., Michaelis, O.E. IV, and Prather, E. S. *J. Nutr.* 111: 1045–1057 (1981).
30. Arky, R. A., in *Present Knowledge in Nutrition.* Nutrition Foundation, Washington, DC, pp. 757–771 (1984).
31. National Diabetes Data Group International Work Group. "Classification and Diagnosis of Diabetes Mellitus and Other Categories of Glucose Intolerance." *Diabetes* 28: 1039–1057 (1979).
32. Keys, A. *Atherosclerosis* 14: 193–202 (1971).
33. Scholfield, D.J., Reiser, S., Powell, A.W., Pankaja, P., and Canary, J.J. *FASEB J.* 2: 5264 (1988).
34. American Council on Science and Health. *Sugars and Your Health*. Summit, NJ, pp. 1–32 (1986).
35. Truswell, A. S. *Naeringsforskning* 17: 12–17 (1973).
36. Yudkin, J. *Nature* 239: 197–199 (1972).
37. Yudkin, J. *Proc. Nutr. Soc.* 31: 331–337 (1972).
38. Walker, A.R.P. *Atherosclerosis* 14: 137–152 (1971).
39. Truswell, A. S. *Food Tech. in Australia* 39: 134–140 (1987).
40. Grande, F. in *Sugars in Nutrition*, ed. by Sipple, H.L. and McNutt, K.W. Academic Press, New York. pp. 401–437 (1974).
41. Grande, F. *World Rev. Nutr. Diet.* 22: 248–269 (1975).
42. Garcia-Palmieri, M.R., Salie, P., Tillotson, J., Costas, R. Jr., , Cordero, E., and Rodriguez, M. *Am. J. Clin. Nutr.* 33: 1818–1827 (1980).
43. Rapoport, J.L. *J. Psych. Res.* 17: 187–191 (1982/83).
44. Kruesi, M. J. P. *Food Technol.* 40: 150–152 (1986).
45. Rapoport, J.L. *Nutrition Reviews* 44: 158–162 (1986).

46. Kruesi, M. J., Rapoport, J.L., Cummings, E.M., Berg, C.J., Ismond, D.R., Flament, M., Yanow, M., and Zahn-Waxler, C. *Am. J. Psych.* 144: 1487–1490 (1987).
47. Behar, D., Rapoport, J.L., Adams, A.J., Berg, C.J., and Cornblath, M. *Nutrition and Behavior* 1: 277–288 (1984).
48. Kruesi, M.J.P. and Rapoport, J.L. *Ann. Rev. Nutr.* 6: 113–30 (1986).
49. Barling, J., and Bullen, G. *J. Gen. Psych.* 146: 117–123 (1985).
50. Fernstrom, J.D. and Wurtman, R.J. *Science* 174: 1023–1025 (1971).
51. Fernstrom, J.D. and Wurtman, R.J. *Science* 178: 414–416 (1972).
52. Li, E.T.S. and Anderson, G.H. *Nutr. Res.* 7: 1329–1339 (1987).
53. Lyons, P.M. and Truswell, A.S. *Am. J. Clin. Nutr.* 47: 433–439 (1988).
54. Ravussin E., Lillioja, S., Knowles, W.C., Christin, L., Freymond, D., Abbott, W.G.H., Boyce, V., Howard, B.V., and Bogardus, C. *N. Engl. J. Med.* 318: 467–472 (1988).
55. Wurtman, J.J. *J. Am. Diet. Assoc.* 84: 1005–1007, (1984).
56. Wurtman, J., Wurtman, R., Mark, S., Tsay, R., Gilbert, W., and Growdon, J. *Int. J. Eating Disorders* 4: 89–99 (1985).
57. Wurtman, J.J., Wurtman, R.J., Growdon, J.H., Henry, P., Lipscomb, A., and Zeisel, S.H. *Int. J. Eating Disorders* 1:2–15 (1981).
58. Danforth, E. *Am. J. Clin. Nutr.* 41: 89–99 (1985).
59. Sims, E.A.H., Danforth, E., Horton, E.S., Bray, G.A., Glennon, J.A., and Salans, L.B. *Rec. Prog. Horm. Res.* 29: 457–464 (1973).
60. Bjorntorp, P. and Sjostrom, L. *Metabolism* 27: 1853–1865 (1978).
61. Glinsmann, W. H., Irausquin, H. and Park, Y. K. *J. Nutr.* 116 No. 11S: S1–S216 (1986).
62. *The Surgeon General's Report on Nutrition and Health.* Government Printing Office, Washington, DC (1988).
63. FDA, Federal Register 53(215), 44862–44876 (1988).
64. Bierman, E.L. *Amer. J. Clin. Nutr.* 32: 2644–2647; 2712–2722 (1979).
65. *Report of the British Nutrition Foundation's Task Force: Sugars and Syrups.* BNF, London (1987).
66. "Position of the American Dietetic Association: Appropriate Use of Nutritive and Non-nutritive Sweeteners." *J. Amer. Dietetic Assn.* 87:1689–1694 (1987).

8

Sugar in Confectionery

C. Michael Barnett*

The nexus of sugar (sucrose) and confectionery products is inescapable. From the simplest rock candy to the most sophisticated confection, the single most important ingredient is sugar. Even with today's technical ability to synthesize nearly any natural compound and to manufacture "new" synthetic materials, no suitable substitute for sugar has been found. Nothing yet devised by humans or nature has the unique sweetening, bulking, and manufacturing properties of natural sugar.

This is not the case for many of the other ingredients used in confectionery manufacture. There exist both natural and artificial, nutritive as well as nonnutritive, substitutes for the corn, egg, dairy, vegetable, flavoring, and coloring materials used in nearly all confectionery products. But there is no substitute for sugar.

So it is appropriate that we begin a discussion of confectionery with a description of the nature of sugar and those special properties that make it particularly suited to the manufacture of confections.

Sugar Properties Important to Confectionery

Sugar (sucrose), the white crystalline material confectioners use daily, is chemically a disaccharide. It is a combined form of two simple sugar molecules: one molecule of glucose and one molecule of fructose. As such, sugar has certain physical and chemical

*C. Michael Barnett has extensive background in the candy industry as a consultant to manufacturers. This chapter is written in memory of the author's father and teacher, Claude D. Barnett, who had over 60 years of experience in the candy industry.

properties that must be understood if it is to be used properly in the manufacture of confections.

Solubility and Boiling Point

One of the most important physical properties of sugar is its solubility in water. Much of candy-making involves the dissolving and subsequent recrystallization of sugar. In fact, historically, candy-making was called the art of sugar boiling.

Under standard atmospheric conditions of temperature and pressure—68°F (20°C) and 29.92 inches of mercury (1014 millibars of pressure)—2 pounds of sugar will dissolve completely in 1 pound of water. This solution of 66⅔ percent sugar solids and 33⅓ percent water is called a standard solution, and is one form in which sugar can be supplied to the candy manufacturer.

As the temperature of the water in which the sugar is to be dissolved changes, the amount of sugar that the water can dissolve (hold) will also change. For example, when 1 pound of water is heated to 200°F (93.3°C), about 4⅔ pounds of sugar will dissolve in it, resulting in a solution that is roughly 82 percent sugar and 18 percent water. Similarly, when one pound of water is cooled to 33°F (1°C), the water will only hold about 1¾ pounds of sugar, resulting in a solution that is roughly 64 percent sugar and 36 percent water.

Another related property of sugar (and other water-soluble materials) is that, when dissolved in water, the boiling point of the resulting solution changes. For example, under standard atmospheric conditions water will boil at 212°F (100°C). When 2 pounds of sugar are dissolved in 1 pound of water, however, the boiling point of the resulting solution will be approximately 220°F (104°C). For any solution of sugar and water, there will be a corresponding boiling point. Conversely, for any given boiling temperature there will be a corresponding amount of sugar dissolvable in the water. This relationship can be plotted on a graph, as shown in Figure 8-1.

This relationship between the boiling point of a solution and the relative amounts of solids and moisture is central to the manufacture of candy. In order to achieve a desired amount of moisture in a product, the initial solution contains more water than needed and is then boiled (cooked) at a temperature that produces the desired finished moisture.

A vacuum can also be used, in combination with cooking, to remove water from a solution. For each inch that atmospheric pressure is reduced, the corresponding boiling point of water is

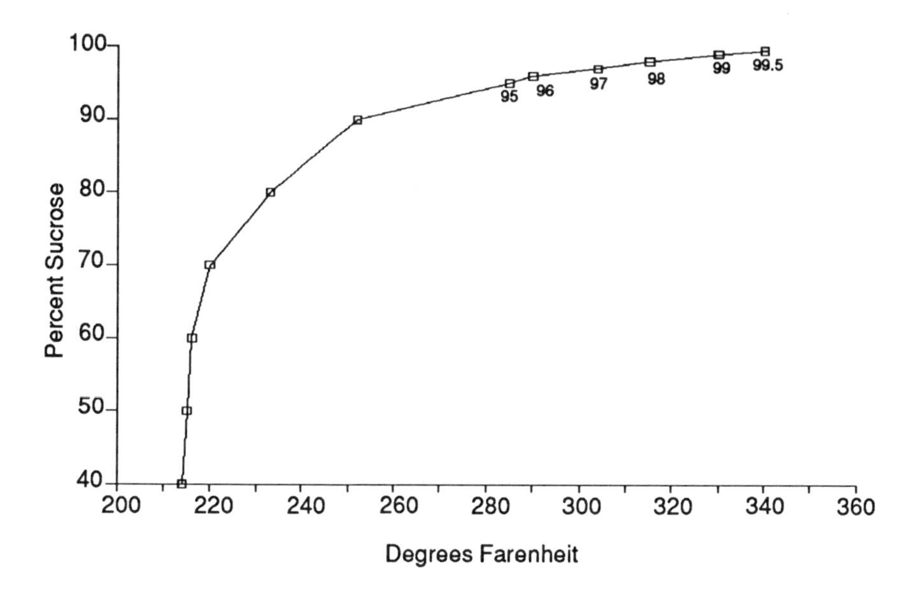

Figure 8-1. Percentage of sugar in solution vs. boiling point.

reduced 2°F. This effect comes into play when a candy factory is located at an elevation higher or lower than sea level. For example, at an elevation of 2000 feet above sea level, which corresponds to a pressure 2 inches lower than that at sea level, water will boil at 208°F, 4° lower than at sea level.

The reduction of boiling temperature through the use of a vacuum is especially useful in products like hard candies, which must have a low-moisture content. To produce a hard candy with a finished moisture content of 2 percent requires atmospheric cooking to 315°F. The same result can be achieved by cooking the batch to 245°F under a 30-inch vacuum, with the added benefits of a shorter production time, lower energy consumption, and a lower batch temperature, which makes it easier and safer to handle. Vacuum cooking is also useful in manufacturing caramels, where the milk and dairy ingredients have a tendency to scorch under atmospheric conditions.

When an 82 percent solution is made by dissolving 4⅔ pounds of sugar in 1 pound of water that has been heated to 200°, an unstable situation called a supersaturated solution is created when the solution is cooled to 68°F. Since water at 68°F will only hold 2 pounds of sugar, and since there are 4⅔ pounds of sugar in an 82 percent solution, there are some 2⅔ pounds of excess sugar in the solution. This excess sugar cannot remain in solution, and so will recrystallize and precipitate from the solution.

The physical treatment given the supersaturated solution as it cools will control how the sugar recrystallizes. If some foreign material like a string or a piece of dried fruit rind is introduced into the solution, and if the solution is allowed to cool slowly with little or no agitation, the excess sugar will recrystallize on the surface of the foreign object, and rock candy or candied fruit peel will be produced. If, on the other hand, the solution is cooled rapidly with agitation, roughly 2⅔ pounds of crystalline sugar will be suspended in 3 pounds of sugar solution (syrup) consisting of 1 pound of water and 2 pounds of sugar—a crude type of fondant.

Graining

In the case of fondant, the size of the crystals precipitated from the supersaturated solution, and the resulting mouthfeel of the product, can vary greatly. Among the factors affecting grain size are (1) the presence, or introduction, of undissolved sugar (seed) crystals in the batch; (2) the temperature at which the cooling batch is begun to be agitated; and (3) the presence of invert sugar in the batch.

Undissolved sugar crystals in a supersaturated solution will act as a catalyst, hastening the precipitation process and providing a pattern for the precipitating crystals. Deliberate seeding (adding controlled amounts of crystals of a known size to a batch) is used to increase fondant production, to produce grain in fudges, caramels, and nougats, and in the processing of after-dinner mints.

Accidental seeding in the production of fondants, mints, and hard candy, on the other hand, is the bane of the candymaker's existence. All residual sugar crystals must be cleaned from the surface of the cooling kettle and its lid. During boiling, the kettle is covered for a short period of time, during which the condensing steam will dissolve any crystals remaining on the lid or kettle rim from the previous batch. Washing any of these crystals back into the present batch will create the potential for undesirable crystal type and graininess. If the surface onto which the cooked candy is poured is contaminated with sugar crystals, the batch will quickly grain into large, undesirable crystals.

The temperature to which a batch is cooled before agitation (creaming) is begun also affects grain size. In general, the higher the creaming temperature, the larger the crystal size will be; and, conversely, the lower the creaming temperature, the smaller the crystal size will be. If the creaming temperature is too high, the crystals will be too coarse, and the candy will taste "gritty." If the creaming temperature is too low, the crystals will be too small, and the candy will lack texture. If the batch is allowed to cool too much, it may become too viscous to beat. The proper creaming temperature for regular fondant is 100°F (38°C) to 130°F (54°C).

We can demonstrate the above principles with the following simple experiment:

INGREDIENTS

1. Sugar	1½ lb
2. Sugar	½ oz
3. Water	8 oz

PROCEDURE

1. In a large cooking pot, combine the sugar [1] and the water [3].
2. Turn on high heat and bring to a boil, stirring gently.
3. Once the mixture is boiling, cover the pot for about 2 minutes. Remove the cover, and use a pastry brush and water to wash any sugar crystals from the inside rim of the pot back into the hot solution.
4. Cook the mixture to 240°F (116°C).
5. Pour one-third of the batch into each of three clean, small fudge trays.
6. Immediately add the sugar [2] to the first tray, and, using a clean 3-inch paint scraper, mix and turn the batch until it becomes opaque and granular.
7. Allow the second tray to cool to 180°F (82°C), and, using a clean 3-inch paint scraper, mix and turn the batch until it becomes opaque and granular.
8. Allow the third tray to cool to 120°F (49°C), and, using a clean 3-inch paint scraper, mix and turn the batch until it becomes opaque and granular.
9. When all three batches have cooled, taste and compare them for texture and mouthfeel.

If the procedure has been followed properly, batch #1 will be coarse, almost like regular granulated sugar, batch #2 will have a somewhat finer grain, and batch #3 will have a fine, smooth texture.

The third factor affecting graining is the amount of invert sugar in the batch. As mentioned before, sucrose is a disaccharide made up of one molecule of glucose and one molecule of fructose that are chemically bonded. Breakage of the bond produces one molecule each of the component monosaccharide sugars, glucose and fructose. This mixture is called invert sugar.

The effect of combining invert sugar in a supersaturated sucrose solution is to reduce the size of the precipitating sucrose crystals. If sufficient invert sugar is added, graining can be postponed or prevented, and a grainless product like hard candy is obtained.

Invert sugar is introduced into a candy batch in one of three ways: (1) the natural inversion of sucrose when it is cooked; (2) the addition of standard invert sugar to the batch; and (3) the addition of invert-containing materials like corn syrup to the batch.

When a sugar solution is cooked, some of the sucrose naturally inverts. The amount of inversion is directly proportional to the length of cooking time. In fact, before the reaction of inversion was understood, candy cooks used to "slack the batch back" (add extra water to the cooking batch) to prolong the cooking time, thereby increasing the amount of invert sugar produced in the batch. The action of cooking time and the production of invert sugar can be demonstrated with the following simple experiment:

INGREDIENTS

1. Sugar	8 oz
2. Water	2 oz
3. Sugar	8 oz
4. Water	6 oz

PROCEDURE

1. In a clean cooking pot, combine the sugar [1] and the water [2].
2. Cook the batch over high heat to 240°F (116° C), covering the boiling batch for about 2 minutes. Make sure to wash any undissolved sugar into the batch with a brush and water.
3. Pour the batch into a clean fudge tray, and allow it to cool to 130°F (54°C).
4. With a clean 3-inch scraper, mix the batch until it becomes opaque and granular.

5. Repeat steps 1 through 4, using the sugar [3] and water [4]. Cook over medium heat to 240°F (116°C).
6. Allow both batches to cool, and taste them for texture and mouthfeel.

If the above procedure is followed properly, batch #2 will have a finer texture than batch #1. Batch #2's longer cooking time resulted in a larger amount of invert sugar being produced. Controlling the amount of invert sugar production in a batch through cooking time and slacking a batch is haphazard at best. It is much better to add the desired amount of invert sugar, and/or other grain-controlling ingredients like corn syrup, to the batch.

The amount of invert produced when a sucrose solution is cooked is also affected by the pH (acidity/alkalinity) of the solution. The lower the pH, the more invert produced when the solution is cooked. Since the principal agent in determining the pH of a sucrose solution is the water used to make the solution, and since the pH of water frequently varies from day-to-day and season-to-season, a prudent candymaker will constantly monitor water pH.

Standardized invert sugar is commercially available for use as an ingredient, or it may be manufactured as follows:

INGREDIENTS

1. Sugar	700 lb
2. Water	300 lb
3. Hydrochloric acid	16 oz
4. Sodium bicarbonate	sufficient to neutralize

PROCEDURE

1. Put water [2] into a steam-jacketed kettle, and warm to 180°F.
2. Gradually add the sugar [1], mixing until it is completely dissolved.
3. Adjust the temperature of the mixture to 160°F.
4. Add the concentrated hydrochloric acid [3], and continue to agitate for 2 hours, maintaining the temperature at 160°F.
5. Neutralize the acid by adding sodium bicarbonate [4] (dissolved in a little water) until the pH of the solution is 5.0.

Note: During the neutralization process, the batch will bubble and foam. Use a kettle large enough to prevent overflowing.

Corn Syrup

Corn syrup is the product of the partial hydrolysis of corn starch. In the hydrolysis process, the long molecular chains of corn starch are broken into sugars such as glucose, fructose, and maltose and into a range of higher-molecular-weight materials called dextrins. The relative amount of glucose, fructose, and other "reducing sugars" in corn syrup is referred to as dextrose equivalent, or D.E. This mixture of sugars is very useful in controlling the recrystallization of sucrose in confectionery manufacture. The dextrins in corn syrup add chewiness or "body" to the candy.

Corn syrups are available with different proportions of reducing sugars and dextrins. The most commonly used corn syrup in general confectionery manufacture is 42 D.E. corn syrup, which contains about 60 percent glucose and other reducing sugars along with about 40 percent dextrins. 64 D.E. corn syrup, which contains about 85 percent reducing sugars and 15 percent dextrins, is primarily used in the manufacture of jellies and gums, where the lower dextrin content translates to more tender products. A special corn syrup with a high percentage of maltose is available for use in hard candy manufacturing.

Basic Candy Types

Some of the basic types of candy are discussed below.

Fondant

Fondant is a suspension of microscopic sugar crystals recrystallized from a supersaturated sugar solution. Fondant is both a confectionery ingredient, used to add grain to caramels, fudges, and nougats and, with the addition of flavoring and texturizing agents, a final confectionery product in the form of "creams."

The amount of moisture in a fondant determines the firmness of the product and ranges from about 15 to 20 percent. Fondants with much less than 15 percent moisture will be dry and hard, while those with more than 20 percent will have enough moisture to support bacterial growth and fermentation. The moisture content of a fondant is primarily controlled by cooking temperature, which can range from 240°F (116°C) to 250°F (121°C).

The actual size of the sugar crystals in a fondant, for proper mouthfeel, ranges from .001 to .002 inch. Crystal size is controlled by the amount of invert sugar developed in cooking or added to the batch, the agitation (creaming) temperature, and the presence of seed crystals.

The process of making fondant consists of cooking a mixture of sugar, water, corn syrup, and sometimes invert sugar to temperature, cooling it to 100°F to 130°F (38°C to 54°C), and then agitating (creaming) it until it recrystallizes (grains).

Cooking can be done in an open-fire kettle, a steam-heated kettle, or vacuum cooking equipment. Cooling and creaming is done on an open beater or in a continuous fondant machine.

Open beaters (also known as Ball beaters or cream beaters) make one batch of fondant at a time. An open beater is a machine with a round, water-cooled bed, usually 6 to 8 feet in diameter and 6 to 8 inches high. Above the bed is a motorized arm equipped with a series of plows and blades. When the motor is turned on, the arm rotates at about 10 RPM, and the plows and blades move through the material on the bed.

In making fondant on an open beater, the cooked fondant syrup is poured on the beater bed and allowed to cool, undisturbed, until reaching the desired creaming temperature (100°F to 130°F). The motor is then turned on, and the plows and blades move through the syrup and agitate it until it grains.

A continuous fondant machine, as the name implies, makes fondant continuously. It consists of a cooling mechanism and a beating section. The cooling mechanism is usually a water-cooled wheel about 10 feet in diameter and 12 to 24 inches wide. The cooked syrup is spread in a thin layer on the turning cooling wheel. The temperature and the speed of the cooling wheel are controlled, so that by the time the syrup reaches the discharge scraper it has cooled to creaming temperature.

The cooled syrup is scraped from the wheel and fed into the beating section. The beating section is a water-jacketed tube about 6 inches in diameter and 3 to 4 feet long. Inside the tube is a motorized shaft with a series of paddles attached in a helical array. When the shaft turns, the paddles agitate the material and push it through the tube toward the discharge end. By controlling the temperature, the speed of the paddles, and the outflow rate, the candymaker lets the product exits the beating section as a continuous flow of finished fondant.

BASIC FONDANT INGREDIENTS

1.	Sugar	80 lb
2.	Corn syrup	20 lb (42 D.E.)
3.	Water	16 lb (2 gal)

PROCEDURE

1. Put the sugar [1], corn syrup [2], and water [3] into a kettle, and mix.
2. Turn on the heat, and bring to a boil with constant stirring. Cover the kettle for about 2 minutes. Remove the cover, and wash any undissolved crystals back into the batch with a brush and water.
3. Cook, without further agitation, to 244°F (118°C).
4. Process the syrup on an open beater or in a continuous fondant machine.
6. Allow the fondant to "mellow" for at least 24 hours before using.

Creams

Creams are essentially flavored fondants, but coloring and other texturizing ingredients are frequently added. Creams generally fall into two categories: plastic or hand-rolled creams (creams shaped by hand or machine) and cast creams (creams deposited in depressions in starch or in rubber molds to create their shape or deposited directly into chocolate shells).

In order to make a cream that is firm enough to be processed yet tender enough to have good eating qualities, invertase is often added to the formula. Invertase is an enzyme that slowly (over a period of up to two weeks) converts some of the sucrose to invert sugar. Because invert sugar is more soluble than sucrose, the proportion of fluid relative to crystalline sugar increases. The increased fluid portion then dissolves some of the sucrose crystals, thereby reducing the crystal size and making the cream finer in texture.

Aerating ingredients like frappé (also known as Mazzetta and egg whip) are often added to creams to make them lighter in texture. Frappé is usually made of egg albumin (egg whites) beaten with corn syrup, although soy albumin and milk albumin are sometimes used in combination with egg albumin.

In commercial operations, the egg albumin is usually dehydrated (spray-dried egg albumin), but fresh or frozen egg albumin may be used. If dehydrated albumin is used, it must be reconstituted by adding the dried egg whites to *cold* water, mixing thoroughly with a wire whip, and allowing it to hydrate for at least 30 minutes. As mentioned before, soy or milk albumin can also be used in frappé but, in general, cannot be used as a complete replacement for egg albumin, since neither of these materials sets when heated, resulting in a loss of air trapped in them.

The proportion of egg, corn syrup, and other ingredients in frappés varies widely, as does the use of frappé in candy formulation. The following frappé formula is representative:

Basic Frappé

Ingredients

1.	Corn syrup	8 lb (42 D.E.)
2.	Sugar	2 lb
3.	Water	½ lb (1 pt to dissolve sugar)
4.	Dry egg albumin	1 lb
5.	Cold water	3 lb (1½ qt to hydrate egg)

Procedure

1. Put the cold water [5] in a stainless steel bowl. Add the dry egg albumin [4], and mix in thoroughly with a stainless steel wire whip. Set aside for 30 minutes.
2. While the egg albumin is soaking, put the water [3] and sugar [2] into a kettle, and cook to 240°F (116°C).
3. Add the corn syrup [1] and mix in. Allow to cool until the temperature is 160°F (71°C).
4. Put the cooked syrup into the bowl of a vertical mixer equipped with a flat beater. Turn the mixer on at low speed, and beat for about 3 minutes.
5. Gradually add the hydrated egg albumin, and mix in.
6. Increase the mixer speed to medium and then high, and beat until the frappé is light and firm (when lifted to a peak with a spatula, the peak will stand by itself).

Plastic or Hand-Rolled Cream.

Plastic or hand-rolled creams are generally made on an open beater. They must be firm enough to be formed and retain their shape throughout the coating process, and then must mellow before eating. They can be flavored with natural or artificial flavors, cocoa, or fruit purée, and colored appropriately.

Open Beater Hand-Rolled Cream

Ingredients

1. Sugar	80 lb
2. Corn syrup	20 lb (42 D.E.)
3. Invert sugar	10 lb
4. Water	16 lb (2 gal)
5. Basic frappé	10 lb
6. Invertase	2½ oz
7. Flavor & color	as desired

Procedure

1. Put the sugar [1], corn syrup [2], invert sugar [3], and water into a kettle, and mix.
2. Turn on the heat, and bring to a boil with constant stirring. Cover the kettle for about 2 minutes. Remove the cover, and wash any undissolved crystals back into the batch with a brush and water.
3. Cook, without further agitation, to 244°F (118°C).
4. Pour the syrup onto a cool cream beater that has been lightly sprinkled with cold water. Sprinkle the top of the batch lightly with cold water, and allow to cool undisturbed to 120°F (49°C).
5. Turn on the beater, and add the invertase [6] and the flavor and color [7].
6. Watch the batch carefully, and as soon as it begins to grain (becomes cloudy), quickly add the frappé [5].
7. Continue beating until the cream is completely grained (it will lose its shine and lump or buck up).
8. Form the creams in an extruder or roll by hand. Coat with chocolate.

Cast Creams.
Cast creams must be fluid enough to flow through the depositing equipment, must then set up firmly enough to be demolded and coated, and then must mellow before eating. Cast creams may be made by simply warming (remelting) fondant to 160°F and depositing it, but they are usually made by combining fondant with a cooked syrup, called a bob. The bob remelts the fondant and serves to increase the amount of candy produced without using additional

fondant. The fondant acts as a seed and, along with the sucrose-invert sugar ratio of the cream and the casting temperature, controls the recrystallization of the bob. Flavoring and coloring are usually added to the cream before it is cast.

Cast Cream

Ingredients

1.	Basic fondant	100 lb
2.	Basic frappé	10 lb
3.	Invertase	2 oz
4.	Color & flavor	as desired
5.	Sugar	30 lb
6.	Corn syrup	20 lb (42 D.E.)
7.	Water	8 lb (1 gal)

Procedure

1. Put the sugar [5], corn syrup [6], and water [7] into a kettle. Turn on the heat, and cook with agitation until the batch begins to boil. Cover the batch and steam for 2 minutes. Remove the cover, and wash any undissolved crystals back into the batch with a brush and water.
2. Cook the batch to 242°F (117°C).
3. Put the fondant [1] into a kettle equipped with an agitator. Turn on low heat and the agitator, and warm the fondant to 100°F (38°C).
4. Add the frappé and mix in slightly.
5. Pour the cooked syrup (the bob) over the fondant/frappé mixture, and mix in.
6. Add the color and flavor [4], and mix in.
7. Adjust the temperature of the batch to 155°F (68°C). Add the invertase, and deposit in impressions in clean, warm (75°F to 90°F; 24°C 32°C), dry (8 percent moisture maximum) starch.
8. When the creams have set (about 24 hours), shake out of starch and coat with chocolate.

The foregoing creams are only two of a nearly infinite variety of candies. Varying the proportions of sugar and corn syrup will change the grain structure and texture of the resulting cream.

Additional ingredients like cream, butter, cocoa, nuts, brown sugar, and maple sugar are only some of the modifications possible.

High-Cooked Candies

The high-cooked family of confections is characterized by a low-moisture content of 1½ to 2 percent, and requires a high cooking temperature of 305° to 335°F (152°C to 168°C). This family of confections includes both the familiar ungrained (clear or translucent) "hard" candies and grained products like after-dinner mints, toffees, and nut brittles. They can be formulated with only sugar (using cream of tartar as a mild acid and sufficient cooking time to produce the invert sugar necessary to control graining), sugar with premanufactured invert sugar, or sugar with some other grain-controlling ingredient, usually corn syrup.

Hard candy can be finished in a variety of ways. The candy itself can be left clear, or it can be pulled on a hook machine that folds air into it and makes it opaque. The candy can be solid, or it can be filled with a center material like jelly, peanut butter, or chocolate paste. And it can be formed into simple or intricate shapes by means of a drop frame (a machine with two engraved rollers, between which the candy mass is passed and shaped), molds formed around sticks to make lollipops, or sized into sticks and crooked to make candy canes.

One popular hard candy that is made with sugar and invert sugar, either as an ingredient or as invert developed in the batch through the use of cream of tartar, is an after-dinner mint or pillow mint. These mints are made as follows:

After-Dinner Mints Using Cream of Tartar

Ingredients

1.	Sugar	80 lb
2.	Water	32 lb (4 gal)
3.	Cream of tartar	¾ oz (20 g)

After-dinner Mints Using Invert Sugar

1.	Sugar	80 lb
2.	Water	32 lb (4 gal)
3.	Invert sugar	12 oz

PROCEDURE

1. Put the sugar [1], water [2], and cream of tartar or invert sugar [3] into a kettle. Turn on heat, and cook over high heat with constant stirring until it begins to boil.
2. Cover the kettle, and allow it to steam for 2 minutes. Remove the cover, and wash any undissolved sugar crystals back into the batch with a clean brush and water.
3. Continue to cook over high heat to 335°F (168°C).
4. Pour the cooked batch onto a cool slab that has been lubricated with a suitable slab dressing like mineral oil.
5. Temper the batch by folding and turning. This is best done by folding the cool sides together and turning the warm side down onto the table. Insert a bar or spatula under the center of the batch, lift it so that the cool sides come together, and turn the folded batch so that a warm side is toward the table. This is continued until the batch is plastic (has a workable consistency). Note: Use only as much folding as is necessary to achieve even cooling. Too much agitation will cause the batch to grain prematurely.
6. Transfer the tempered batch to a pulling machine. Pull for 3 minutes. Add 1 ounce oil of peppermint to the batch while it is being pulled. Color may also be added if desired.
7. Remove the batch from the pulling machine, and run it through a mint cutter.
8. Layer the cut mints in powdered sugar in trays, and put in a room heated to 120°F (49°C) for 12 to 24 hours.
9. Remove the mints from the hot room, and sift to separate the mints from the powdered sugar. The surface of the mints should be grained. Pack the mints as desired. Within a week or two, the whole mint will have grained, and will have a smooth, soft consistency.

Note: Owing to variances in cooking equipment, environmental conditions, elevation, and water pH, cooking time and temperature, cream of tartar/invert sugar amount, and hot-room temperature/holding time may have to be adjusted. If the grain is too coarse, increase the cream of tartar/invert sugar slightly, and/or lower the temperature/reduce the holding time in the hot room. If grain is too fine, or the mints do not grain off at all, decrease the amount of cream of tartar/invert sugar, and/or increase the hot-room temperature/holding time.

BUTTER CRUNCH

Another popular high-cooked candy made with sugar and invert sugar is butter crunch:

INGREDIENTS

1. Sugar	20 lb
2. Invert sugar	1 lb
3. Butter	10 lb
4. Almonds (raw)	10 lb (finely chopped)
5. Lecithin	1 oz
6. Baking soda	1 oz

PROCEDURE

1. Put the sugar [1], invert sugar [2], butter [3], and lecithin [5] into a kettle. Turn on medium heat, and cook with constant stirring until all ingredients have combined.
2. Add the almonds [4], and continue cooking and stirring at 295°F (146°C) until the almonds are lightly roasted.
3. Remove from the fire. Add the baking soda [6], and stir in vigorously.
4. Pour out onto a warm slab, and spread to a uniform thickness of about ¼ inch.
5. Score the batch into bite-size squares. A roll knife (several circular knife blades spaced at one inch intervals on a shaft/handle) is useful for this purpose.
6. Insert a long spatula between the candy and the table, and run it around the edge of the batch to separate it from the table and prevent it from sticking.
7. When the candy has cooled, break it on the score marks and pack as desired, or dip in milk chocolate and roll in finely chopped roasted almonds.

Clear or ungrained candies (what we call hard candy) can also be made with sugar and invert sugar, either developed in the cooking process or added as an ingredient:

All-Sugar Hard Candy Made With Cream of Tartar

Ingredients

1.	Sugar	100 lb
2.	Water	20 lb (2.5 gal)
3.	Cream of tartar	1 oz

All-Sugar Hard Candy Made with Invert Sugar

1.	Sugar	100 lb
2.	Water	20 lb (2.5 gal)
3.	Invert sugar	10 lb

Procedure

1. Put the water [2], sugar [1], and either cream of tartar or invert sugar [3] into a kettle. Cook over high heat with constant agitation until the batch begins to boil.
2. Cover the batch and allow to steam for 2 minutes.
3. Remove the cover, and wash down the sides of the kettle with a brush and water to remove any undissolved sugar crystals.
4. Cook to 335°F (168°C) as rapidly as possible.
5. Remove from stove, and pour onto a cool slab or water-cooled table that has been lubricated with a suitable slab dressing like mineral oil. Allow the batch to "rest" on the slab until the edges begin to harden or crust.
6. Add flavor (and color if desired) by sprinkling onto the middle of the batch. Fold the edges of the batch in toward the center.
7. Temper the batch (and distribute the color and flavor) by folding and turning. Note: Use only as much folding as is necessary to distribute the color and flavor and to achieve even cooling. Too much agitation will cause the batch to grain.
8. Run the batch on the desired forming machinery. All-sugar hard candy is considered to be of a high quality and is quite sweet. It has a relatively short shelf-life, however, and will begin to grain off and become sticky and/or cloudy in appearance within a month or two.

To produce hard candy with a longer shelf-life, candy makers typically use corn syrup in the formulation. Corn syrup is an excellent grain controller and adds "body" or chewiness to the candy as well. The amount of corn syrup added to a hard-candy formula can range from 20 to 50 percent. The percentage of corn syrup used affects the quality, taste, shelf-life, cost, cooking temperature and, in some cases, the cooking method of the hard candy.

Because corn syrup is a less expensive ingredient than either sugar or invert sugar, its use in hard candy reduces the overall cost of the product. Corn syrup is also less sweet than sugar and will reduce the sweetness of the resultant candy. Corn syrup also has different cooking properties than sugar alone. Formulas using corn syrup will have a lower final cooking temperature than all-sugar formulas. Corn syrup also tends to darken as it is cooked, and thus will change the appearance of the finished candy. These effects of corn syrup on quality, shelf-life, and cooking temperature are summarized in Table 8-1.

Table 8-1. Effects of Corn Syrup as an Ingredient in Finished Candy

Formula	Quality	Shelf-life	Final Cook Temp
ALL-SUGAR	Very High	Short	335°F (168°C)
CORN SYRUP:			
20%	High	Med. Short	325°F (163°C)
30%	Med. High	Medium	315°F (157°C)
40%	Medium	Med. Long	310°F (154°C)
50%	Low	Long	305°F (152°C)

Hard candy is often cooked in a vacuum system. In this method, the batch is precooked to a relatively low temperature (240°F to 280°F; 116°C to 138°C) either over an open fire or in a steam-jacketed kettle. The batch is then transferred to a sealed vessel (or in the case of a simpler vacuum kettle, the cover is closed), and a vacuum of 29 to 30 inches is drawn. The vacuum reduces the moisture in the batch to the desired 1½ to 2 percent. Vacuum cooking is the preferred method for cooking hard candies with a formulation containing 30 percent or more corn syrup.

The lower cooking temperatures employed in vacuum cooking reduce the darkening of the corn syrup and make the batch safer to handle than candy cooked in an open kettle. Batches cooked in a vacuum system will also be more viscous than open-kettle cooked candy (owing to the lower temperature), and will not be suitable for molding or depositing. Also, candy cooked in a vacuum system will be slightly more opaque than candy cooked in an open kettle because of air trapped in the candy mass by the vacuuming process.

To produce crystal-clear hard candy for molding or depositing from formulas containing corn syrup, a scraped surface heat exchanger, or "thin film cooker," is used. This is a piece of equipment that passes a very thin layer of candy over a very high temperature tube, "flashing off" the moisture without darkening the corn syrup. Thin-film cookers produce a continuous stream of high temperature candy that is suitable for molding or depositing. They are used principally in high-volume production.

Peanut Brittle

Peanut brittle and other nut brittles are essentially hard candies with nuts and other flavoring agents added. Butter, salt, and vanilla are typically used. In some formulas, baking soda is added when cooking is complete to puff up (aerate) the batch and make it lighter.

High-Quality Old-fashioned Peanut Brittle

Ingredients

1.	Sugar	20 lb
2.	Corn syrup	16 lb (42 D.E.)
3.	Invert sugar	1 lb
4.	Water	4 lb (2 qt)
5.	Peanuts (raw)	16 lb
6.	Butter	1 lb
7.	Natural vanilla	2 oz
8.	Fine salt	6 oz
9.	Baking soda	2 oz

Procedure

1. Put water [4], sugar [1], corn syrup [2], and invert sugar [3] into a kettle (large enough to hold the batch and allow room for it to puff up). Turn on high heat; cook with agitation until the batch begins to boil.
2. Cover the batch, and allow it to steam for 2 minutes. Remove the cover, and wash any undissolved sugar crystals back into the batch with a clean brush and water.
3. Cook to 240°F (116°C), and then add the peanuts [5]. Continue cooking, without stirring, until the batch reaches 295°F (146°C).

4. Resume stirring and continue cooking the batch until the peanuts are golden brown—about 305°F (152°C).
5. Remove the kettle from the fire, and place it on a kettle stand. Add the butter [6] and vanilla [7], and mix.
6. Combine the fine salt [8] and baking soda [9] in a wire mesh strainer. Sift into the batch while it is being vigorously stirred.
7. Pour the batch out onto a cool slab that has been lubricated with a suitable slab preparation like mineral oil. Quickly spread out the batch so that it is about 3⁄8-inch (1 nut) thick.
8. Allow the batch to cool for a few minutes. Cut it into manageable-sized pieces, and turn them over.
9. Pull and stretch the pieces to thin out the candy between the nuts. (Wear clean cotton work gloves for this operation.)
10. Allow the brittle to cool completely. Break up into desired size pieces, and pack as desired.

Note: Old-fashioned nut brittles are classically made in copper kettles. There is a catalytic chemical reaction that takes place when fats are heated in the presence of copper. This reaction gives peanut and other brittles a characteristic flavor. However, the reaction also accelerates the rate at which the fat will oxidize (go rancid). If copper kettles are used, the addition of an antioxidant like BHT or BHA is recommended.
Note: The cost and the quality of nut brittles are primarily controlled by adjusting the ratio of nuts to candy and by the type and amount of fat and flavor used. For example, the butter can be partially or completely replaced with margarine or vegetable fat.

Chewy Candies

Chewy candies are similar in formulation to the ungrained hard candies discussed earlier, with some significant differences. These candies have a higher percentage of corn syrup and are cooked to a lower temperature than hard candy, and therefore have a higher moisture content and a softer consistency. The chewy consistency of these candies is sometimes enhanced by aerating. Air is incorporated either by folding the candy on a pulling machine (as in taffies and kisses) or by including an aerating ingredient like beaten egg whites, as in nougats. Chewy candies also require some kind of fat to make them machinable and to keep them from sticking to the consumer's teeth. The handling of the fat is crucial to the successful manufacture of aerated candies.

The following molasses kiss is typical of chewy candies that are aerated by pulling:

OLD-FASHIONED MOLASSES KISS

INGREDIENTS

1. Sugar	30 lb
2. Corn syrup	60 lb (42 D.E.)
3. Molasses	10 lb
4. 76° Vegetable fat	2½ lb
5. Butter	½ lb

PROCEDURE

1. Put the sugar [1], corn syrup [2], and vegetable fat [4] into a kettle. Turn on heat and cook to 255°F (124°C).
2. Add the butter [5] and molasses [3], and continue cooking to 262°F (128°C).
3. Remove from the heat, and pour onto a cool slab that has been lubricated with suitable slab dressing such as mineral oil. Allow to cool.
4. Transfer the cooled batch to a pulling machine, and pull for three minutes. (Note: one ounce oil of peppermint may be added to batch while pulling for a molasses-peppermint kiss.)
5. Run the batch out on a cut-and-wrap machine. Pack as desired.

Nougat

Another very common method of aerating chewy candies is through the incorporation of frappé. Nougats are typical of chewy candies aerated with frappé.

The following nougat is a basic one. Other variations include the addition of fondant to give the nougat a grained texture, the inclusion of gum drops in the batch, and shaped colored centers (like red hearts) giving the finished product a "picture" within the candy.

BASIC NOUGAT

INGREDIENTS

1.	Sugar	25 lb
2.	Corn syrup	25 lb (42 D.E.)
3.	Invert sugar	6 lb
4.	Water	5 lb
5.	Basic frappé	15 lb
6.	98° Vegetable fat	2½ lb
7.	Salt	2½ oz
8.	Chopped nuts	4 lb (optional)
9.	Flavor (vanilla or maple)	as desired

PROCEDURE

1. Put the sugar [1], corn syrup [2], invert sugar [3], and water [4] into a kettle.
2. Turn on heat and cook to 275°F (135°C). Steam the kettle for two minutes, and wash any undissolved sugar crystals back into the batch. Remove from heat, and allow the batch to cool to 250°F (121°C).
3. Put the frappé [5] into the bowl of a vertical mixer equipped with a flat beater. Turn the mixer on to low speed, and gradually pour the cooked portion into the mixer.
4. Add the salt [7], flavor [9], and nuts [8], and mix until blended.
5. Turn the mixer on high speed, and add the vegetable fat [6]. Mix in as quickly as possible.
6. Pour the batch out onto a slab equipped with ¾- inch bars, and allow to cool.
7. Cut the nougat to desired size, and wrap or dip in chocolate, as desired.

Note: It is extremely important that the batch not be mixed any more than necessary to distribute the vegetable fat after it has been added. The fat reduces the surface tension of the air bubbles in the frappé, and the batch will go flat if mixed too much.

Caramels

Caramels are a unique member of the chewy candy family. While caramel formulations are similar to other chewy candies, they also necessarily contain some form of milk protein. Whole milk is not generally used, as it contains too much water and results in extremely long cooking times and too much invert production. Evaporated or condensed milks are preferred. Caramels are handled and packaged in a variety of ways. They can be ungrained, or fondant can be added to produce a grained caramel. They can be made in slabs, cut, and wrapped, or they can be machined in cut-and-wrap equipment.

Caramel can be deposited in starch molds, on a bed of nuts (to make turtles), or rolled in nuts to make pecan logs. Caramel can also be layered with nougat, marshmallow, and other candies to make a wide variety of products.

Higher-quality caramels will be made with butter, cream, and sweetened condensed whole milk, while less expensive caramels will be made with vegetable oils and condensed skim milk or milk substitutes containing milk protein.

Milk proteins are required not only for the flavor they impart to caramel but also for the characteristic caramel color and texture. When milk protein is cooked in the presence of sugar, it undergoes a chemical reaction called the Maillard reaction. This reaction is characterized by the darkening or "browning" of the milk protein. Additionally, the reaction causes a thickening of the milk protein (akin to the setting of egg whites when they are cooked) that enables caramels to stand up (retain their shape) while still being soft and chewy.

Cooking caramels can be a tricky business. They can be cooked in a kettle over an open fire, in steam kettles, and in vacuum kettles.

The classic method of cooking caramel is in a copper kettle over an open fire and is still used in some gourmet and small production shops today. This method is limited to small batches and requires a high level of skill and constant vigilance to insure that the milk in the batch does not scorch. Cooking caramel over an open fire involves cooking the nondairy ingredients (sugar, corn syrup, and invert sugar) to about 240°F (116°C). The milk is then added to the boiling batch in a thin stream so that the batch never stops boiling. When all the milk has been added, the batch is cooked to its final temperature of 242°F to 248°F (117°C to 120°C). The butter and flavorings are then added.

A much easier and faster method of cooking caramels is in a steam-jacketed kettle. Because the temperature of steam never exceeds 300°F (149°C), the problem of scorching the milk is much reduced. Additionally, since the steam heats nearly all of the kettle (not just the bottom as a gas burner does), the batch generally cooks more quickly. With a steam kettle, all of the ingredients (except the flavor) can be put in the kettle at once and cooked together. For higher production, special high-speed caramel steam kettles are available. These kettles have very high-speed agitators equipped with precision scrapers that run very close to the sides of the kettle. The agitators literally fling a thin film of caramel up onto the walls of the kettle and then scrape it off again before it has a chance to scorch—something like a thin-film hard-candy cooker.

Caramel can also be vacuum-cooked. In this process, the ingredients are cooked to about 238°F (114°C) and then subjected to a 15-inch vacuum for 5 minutes. Vacuumed caramels are usually lighter in color than steam or open-fire caramels, since the cooking time is shorter and the temperature is lower. Additionally, nonvolatile flavors must be used, or flavors must be added after vacuuming, as the vacuum will remove volatile flavors.

The following formula is for a high-quality caramel made in a steam kettle:

High-Quality Caramel

Ingredients

1. Sugar	20 lb
2. Corn syrup	20 lb (42 D.E.)
3. Invert sugar	5 lb
4. Cream	2 lb
5. 92° Vegetable oil	2 lb
6. Evaporated milk	30 lb (unsweetened)
7. Butter	2 lb
8. Fine salt	6 oz
9. Baking soda	½ oz
10. Flavor	as desired

Procedure

1. Put the sugar [1], corn syrup [2], invert sugar [3], cream [4], and vegetable oil [5] into a steam-jacketed kettle equipped with an agitator and side scrapers.
2. Mix the ingredients for about 5 minutes, and then turn on the heat. Cook to 240°F (116°C).
3. Add the baking soda [9] to the evaporated milk [6], and add to the batch. Cook to 242°F (117°C).
4. Add the butter [7] and salt [8], and mix in well.
5. Add the flavor [10] and mix in.

6. Pour the batch out on a slab between ¾- inch bars, and allow to cool.
7. Cut the caramel to size on a caramel cutter or process on a cut-and-wrap machine.

Note: Evaporated milk will sometimes curdle when cooked, especially in an acidic environment. The addition of a small amount of baking soda will prevent this.

Fudge

The story goes that a famous Austrian candymaker and his helper were making a batch of caramel one day. The batch was nearly finished when one of the candymaker's best customers entered the shop. She wanted to discuss the confectionery arrangements for a grand party she was planning. Not wanting to offend the lady by asking her to wait until he had finished the caramel, the candymaker told his assistant to continue stirring the batch while he spoke with the customer.

The conversation apparently took longer than the candymaker had expected. When he returned to his helper, he found him struggling, dutifully stirring the batch of caramel, which, by this time, had cooled and lumped into something more resembling fondant than caramel.

"You idiot! It is ruined," shouted the candymaker, looking at his once-beautiful caramel. "I will have to throw it out and begin again!"

Being a frugal sort, however, the candymaker ventured to taste the "ruined" batch before throwing it out. To his surprise, it was wonderful! He decided to market his new confectionery discovery, naming it after his helper, who had serendipitously created it: Gustav Fudge.

Whether or not this is truly the origin of fudge, this story does accurately describe the basic structure of fudge. It is primarily a grained caramel. Like the Austrian candymaker, today's candymakers could achieve this by agitating a cooked caramel until it grains. It is much easier, however, to grain the fudge by the addition of fondant.

Like caramel, fudge contains milk protein and fat, as well as sugar and a grain controller like invert sugar or corn syrup. As for fondants and creams, the manufacturing procedures and temperatures control the crystal size and the texture of the finished product.

In general, the most critical step in making fudge is the temperature of the batch when the fondant is added. If this temperature is too high (above 125°F), the texture of the fudge will be coarse and gritty. If the temperature is too low, (below 90°F), the fudge may not grain at all.

Vanilla Fudge

Ingredients

1. Sugar	30 lb
2. Corn syrup	20 lb (42 D.E.)
3. Cream	20 lb (20 percent butterfat)
4. Butter	5 lb
5. Salt	5 oz
6. Basic fondant	10 lb
7. Vanilla	4 oz

Procedure

1. Put the sugar [1], corn syrup [2], cream [3], and butter [4] into a steam kettle equipped with an agitator and side scrapers. Mix thoroughly.
2. Turn on the heat and cook to 230°F (110°C).
3. Transfer to another kettle equipped with an agitator to cool. Allow the batch to cool to 125°F (52°C) without agitation.
4. Add the salt [5], fondant [6], and vanilla [7]. Turn on the agitator, and mix until the batch just begins to grain.
5. Put the batch out in fudge pans, or pour on a table between 3⁄4-inch bars, and allow to set.
6. Score and cut to desired size. Package as desired.

Note: Once this fudge begins to grain, it will set quickly. If the batch should set before it is all put out, it can be remelted in a double boiler or a cream breaker. The remelt temperature must be kept between 145°F and 155°F (63°C and 68°C). Fudge can be made up in bulk and then remelted and put out as needed.

References

1. Barnett, C. D. *Candy Making as a Science and an Art*. Don Gussow Publications, New York, NY. 1960.
2. Barnett, C. D. *The Science and Art of Candy Manufacturing*. Books for Industry, New York, NY. 1978.

3. Richmond, W. L. *Candy Production—Methods and Formulas.* Manufacturing Confectioner, Chicago. 1948.
4. *Twenty Years of Confectionery and Chocolate Progress.* AVI/Van Nostrand Reinhold, New York, NY. 1970.
5. Minifie, B.W. *Chocolate, Cocoa, and Confectionery Science and Technology.* 2nd ed. AVI/Van Nostrand Reinhold, New York, NY. 1980.
6. U.S. Food & Drug Administration. *Current Good Practice in Manufacturing, Processing, Packing, or Holding Human Food,* 42 Federal Register 14350, March 15, 1977.

9

Sugar in Bakery Foods

J. G. Ponte, Jr.*

Sweeteners are among the most important ingredients in bakery foods and, in this context, have a long history of usage. The importance of sweeteners to the baking industry is due to the large quantities used and to the vital functions these materials perform in a wide array of bakery products. The contribution of sweeteners to sweetness and flavor in these foods is well recognized. Perhaps less well known are other important functions of sweeteners, including crumb tenderizing, textural effects, moisture retention, formation of crust color, shelf-life improvement, preservative action, fermentation substrate, whipping and creaming aid, and bulk support. Some of these functions will be dealt with in the following discussion.

Although a number of sweeteners may be utilized in the production of various bakery foods, only sucrose will be considered in this chapter. Perhaps no sweetener is as versatile as sucrose in baking applications. Sucrose (sugar) has a high level of sweetness and can be used in both crystalline and liquid forms (usually 67 percent solids) in bulk systems. Thus, unlike some other sweetening systems, sucrose can be readily used to produce virtually any bakery food, including yeast-raised products, chemically leavened products, and icings and fillings.

Sucrose Products in Baking

Bakery technologists have used a wide range of sucrose products for various applications. Granulated sugar is available in many

*J. G. Ponte, Jr., is Professor, Department of Grain Sciences and Industry, Kansas State University, Manhattan, KS.

particle sizes ranging from very fine to coarse and include extra fine, fine granulated, bakers special, industrial fine, manufacturers granulated, and sanding sugars. The baking application determines the sugar type. For example, some fillings in which sugar is not fully dissolved are prepared with fine granulated sugar to avoid grittiness. Sanding sugars, on the other hand, are used as toppings on certain cakes and cookies for which residual granules on the baked surface are desirable for purposes of appearance, flavor, and texture. The most commonly used form of sucrose is bakers special granulated sugar, a medium-sized material.

Powdered sugars are available also in various particle sizes commonly ranging from 2X to 12X, with the higher numbers designating decreasing particle size. Powdered sugars are used in a number of applications. Some baked cookies and cakes receive a dusting of powdered sugar for appearance and flavor. Some cake systems such as angel food, in which flour and sugar are carefully folded into the batter during the later mixing stages, use powdered sugars. Powdered sugars are widely used in icings and fillings to avoid grittiness that would result from the use of coarser sugars. To obtain free-flowing characteristics, 3 percent cornstarch or 1 percent or less tricalcium phosphate is added to powdered sugars during manufacture.

Fondants are prepared with sucrose of very fine particle size for use in icings or fillings where an especially smooth mouthfeel is sought. Paste fondants are made by controlled crystallization, with agitation, of sucrose solutions, whereas dry fondants are prepared by blending finely ground sucrose with invert sugar or maltodextrin.

Invert sugar is prepared by treating sucrose with enzymes or mild acid, which releases free glucose and fructose. Several invert sugar grades are available, varying in the degree of sucrose inversion. Invert sugars have humectant properties and are used in bakery foods to retain moisture and prolong shelf-life. In icings, invert sugar imparts gloss and improves pliability.

Other forms of sucrose available to the baker include brown sugar and molasses. Various grades of brown sugar are available, ranging from light to dark brown in color. Molasses is also available in numerous grades varying in color and flavor, depending on the method of production and how much sucrose is removed during processing. Both brown sugars and molasses are used for their contribution to flavor and color in a wide variety of bakery foods, including breads, cakes, and cookies. In recent years, emphasis on "natural" foods has caused increased production of the darker types of breads and sweet goods, which often feature the inclusion of brown sugars and molasses.

More information on the characteristics of various sucrose products may be found elsewhere in this book.

To place the use of sucrose in bakery foods in perspective, it is useful to recognize that the baking industry is an important part of the U.S. food industry. The baking industry had sales of nearly $50 billion in 1989 (1), distributed among four major industry segments: wholesale, 56.4 percent; retail, 10.6 percent; in-store, 15.5 percent; and food service, 17.5 percent. The largest industry segment, the wholesalers, had a product distribution in 1988 (2) of bread and rolls, 40 percent; cookies and crackers, 35 percent; various sweet goods, 17 percent; and frozen bakery foods, 8 percent. Sugar consumption by the wholesale baking segment alone in 1982 was something over 758,000 tons (3).

Yeast-Leavened Bakery Foods

Usage of sugar in yeast-leavened products can range from about 1 to 11 percent (based on flour) in breads and up to about 20 percent in Danish sweet goods. For example, "lean" breads, such as French and Italian hearth breads, are typically made with lower sugar levels, whereas premium white breads may be made with 11 percent or more sugar to obtain desired flavor and texture characteristics. Standard U.S. white pan bread, which is the single most important bakery food in terms of tonnage produced, is formulated with about 8.5 percent sugar.

Specific functions of sugar in yeast-leavened bakery foods include providing yeast fermentables, enhancing flavor, improving crust color and toasting properties, modifying crumb texture, and extending shelf-life. In terms of flavor, it is of interest that sugar usage in white pan bread has crept up over the years, from about 2 percent in the 1920s to an average of about 8.5 percent today. U.S. consumers evidently desire the flavor and functionality imparted to bread by sugar.

Breadmaking Processes

Bread in the U.S. is made by a number of processes, but the three most predominant procedures are the sponge dough, liquid ferment, and short-time (sometimes called "no-time") methods. The sponge dough method is used most widely, accounting for 60 percent of the bread produced today. Liquid ferment is the second most important method and is gaining in popularity. Both sponge dough and liquid-ferment methods are preferred by the large wholesale bakers. The short-time method is utilized by a few of the larger bakers and by most retail operators.

Table 9-1 summarizes formulations and procedures for making white pan bread by these three methods. Further details are available from a number of sources (e.g., 4, 5, 6). Information on producing other types of breads is also available (7).

Table 9-1. Comparison of Breadmaking Procedures

	Sponge Dough		*Liquid Ferment*		*Short Time*
Ingredient	SPONGE	DOUGH	FERMENT	DOUGH	
			parts by weight		
Flour	65.0	35.0	30.0	70.0	100.0
Water	37.0	27.0	50.0	13.5	65.0
Yeast	2.5	—	3.0	—	3.5
Salt	—	2.0	0.5	1.5	2.0
Sugar	—	8.5	1.0	7.5	8.5
Shortening	—	3.0	—	3.0	3.0
Yeast food	0.5	—	0.5	—	0.5
Dairy ingred.	—	2.0	—	2.0	2.0
Emulsifier	—	0.5	—	0.5	0.5

Procedures

SPONGE DOUGH

Sponge. Sponge mixed 3 minutes at 77°F; ferment 4 hours at 86°F.

Dough. Mix fermented sponge and dough ingredients for 10 to 12 minutes at 80°F. Rest dough for 15 to 20 minutes. Divide dough into desired-size pieces, round, and allow 10-minute rest period. Sheet, shape, and pan. Pan proof dough pieces for 60 minutes at 107°F, 85% RH. Bake for 18 minutes at 445°F.

LIQUID FERMENT DOUGH

The liquid ferment stage is fermented 2 hours at 88°F. Add to dough ingredients; dough is mixed for 13 to 14 minutes and then processed as in sponge dough procedure.

SHORT-TIME DOUGH

All ingredients are mixed at once for 13 to 14 minutes. Dough is rested for about 10 minutes, then processed as above. In addition to tabulated ingredients, short-time doughs may contain additional oxidants, reducing agents, and protease enzymes to enhance development.

Role of Sugar in Fermentation

An important function of sugar in dough, as previously noted, is to support yeast fermentation.

Bakers yeast, *Saccharomyces cerevisiae*, gives off carbon dioxide gas during metabolism of simple sugars, causing the leavening action that is important in the production of porous, palatable bread. The yeast utilizes sugars obtained from flour or added as part of the dough formula (5).

During fermentation of the first stage (sponge) in the sponge dough process, a sequence of events occurs that is depicted in Figure 9-1. At the beginning of fermentation, carbon dioxide is rapidly produced as the yeast uses the small amount of free sugars naturally present in flour. After about one hour, the rate of carbon dioxide production drops because the free sugars are exhausted. At this

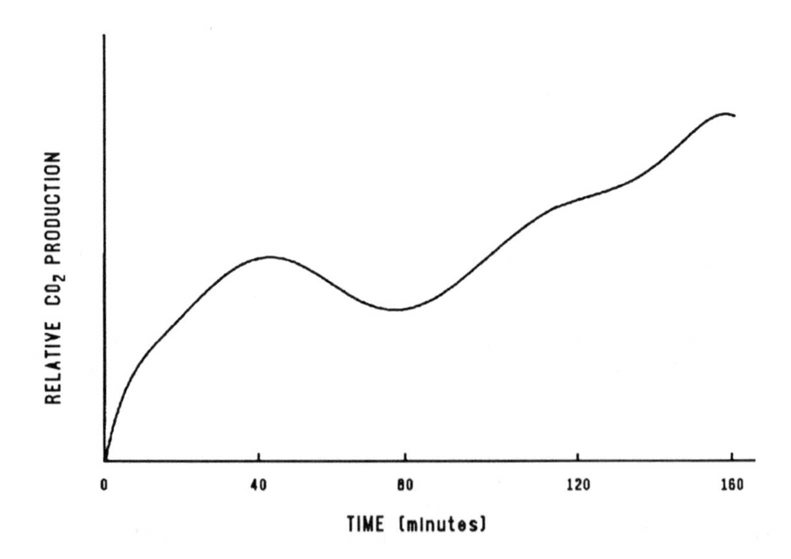

Figure 9-1. Carbon dioxide production during fermentation.

time, the yeast adapts to the fermentation of maltose, and gas production again picks up. The maltose results from amylase hydrolysis of wheat starch that was physically damaged during the flour milling process. After about three hours, the maltose becomes depleted, and gas production tapers off.

During the dough stage of the sponge dough process, or in any breadmaking process in which all formula ingredients are combined, the bakers yeast will metabolize some of the sugar added as an ingredient. The sugar incorporated in the dough is almost immediately hydrolyzed into its constituent monosaccharides, glucose and fructose, by the action of the enzyme invertase, present in bakers yeast. The yeast then will begin to ferment both simple sugars, but at differing rates. Glucose is preferred and will disappear at a faster rate than fructose. Figure 9-2 illustrates the fate of these sugars.

In bread made with sucrose, residual fructose will be present at higher levels than residual glucose. For example, Ponte and Reed (5) reported that sponge dough bread made with 6.7 percent sucrose contains 2.6 percent residual fructose and 1.7 percent residual glucose (both on dry solids basis). Thus, roughly 2.5 percent of the added sugar is utilized by the yeast during the breadmaking process.

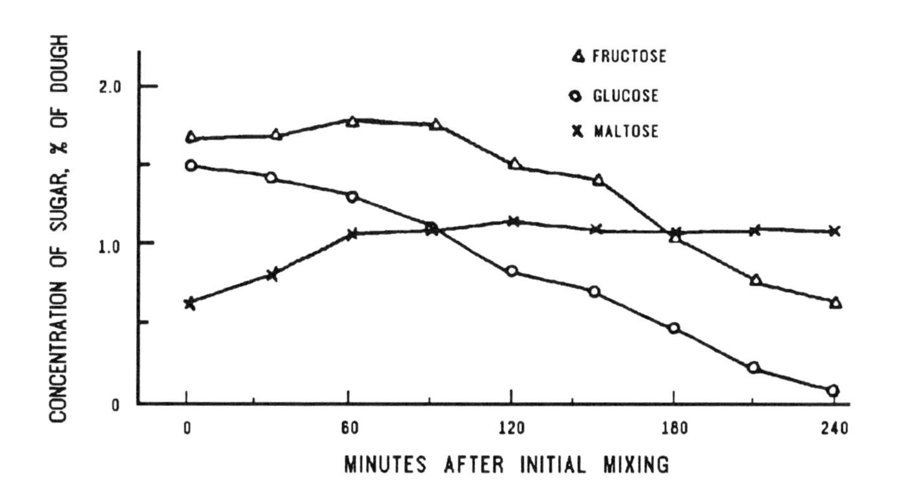

Figure 9-2. Fermentation rate of simple sugars by bakers yeast in bread dough (8).

Role of Sugar in Bread Flavor

The flavor of fresh bread is universally appealing and undoubtedly accounts for much of the widespread acceptance of bread in diets around the world. Bread flavor is subtle and has been elusive to characterize, in spite of much effort (9). However, bread flavor is generally recognized as emanating from two principal sources, both involving sugar: fermentation and crust browning.

The characteristic flavor of yeast-leavened bread partially develops during yeast fermentation of simple sugars into an array of compounds that subsequently react with other dough components during the baking stage. During baking, some of the flavor compounds escape, while others interact with amino acids, and perhaps other materials, to produce characteristic bread flavor. Many dough fermentation by-products have been identified (10), including classes of organic esters and acids, alcohols, and carbonyl compounds. The complexity of bread flavor is indicated by the fact that attempts to synthesize bread flavor by combining organic compounds identified in bread, liquid ferments, or dough have not been successful.

Although the precise role of yeast fermentation of sugar in bread flavor development is far from clear, it seems certain that fermentation by-products are important. Even in the short-time dough process, in which traditional bulk fermentation is not used, the dough pieces must be proofed prior to baking. During this stage and into the early part of baking, the yeast actively produces by-products, which contribute to flavor either directly or via flavor precursors. Breakdown products resulting from the action of yeast and flour proteases on flour proteins also play a role in flavor and color developments (5). Some of these products, including peptones and polypeptides, react with sugars to produce desirable flavors during baking.

Sugar and Crust Browning

Crust browning, as previously noted, is also important in bread flavor. Some of the activities related to yeast fermentation also influence the extent of crust browning. The importance of crust browning to bread flavor may be shown by baking bread in a microwave oven in which browning will not take place, or by removing the crust of conventionally baked fresh bread and storing it until cooled. In both instances, the breads will not exhibit characteristic bread flavor.

Two different thermal browning reactions are responsible for crust browning, caramelization and the Maillard reaction. Both reactions are nonenzymatic and require heat, with caramelization requiring far more energy than the Maillard reaction.

Under heat, caramelization transforms sugars from colorless, sweet compounds into substances ranging in color from pale yellow to dark brown and in flavor from mild, caramel-type to burnt and bitter. This transformation involves a sequence of complex reactions (11).

The Maillard reaction involves interactions between free reducing sugars and the free amino groups of amino acids, peptides, or proteins, leading to melanoidin formation. This is a complex reaction, and recent work (12) has suggested new mechanisms to account for it. Lane and Nursten (13) studied eight sugars and twenty-one amino acids heated under various conditions and found typical bakery-food aromas when glucose was heated with a number of amino acids at a one-to-one ratio under certain conditions of time and temperature.

The extent of crust browning, of course, is related to the amount of sugar in the dough system. Bread made with small quantities of sugar tends to be pale or light-golden in color, whereas bread made with higher sugar levels has a darker, golden-brown color. Toast made from these breads follow similar patterns; bread with higher sugar levels requires less toasting time to achieve a given color level than bread made with little sugar.

Crumb Texture and Firming

Crumb texture may be modified by the presence of sugar in the dough formulation. Two mechanisms are involved in this modification, gluten hydration and starch swelling.

During the mixing of dough, an early event involves hydrating dough components. Gluten, the protein-lipid complex that constitutes the structural backbone of dough, is formed only in the presence of water and mixing energy. As sugar levels in dough are increased, more water is rendered less available for gluten hydration. A consequence of this is that longer mixing times are required to optimally mix, or "develop," the gluten. This effect of sugar delaying dough development was shown some years ago (14). A consequence of inhibiting gluten development is that the bread crumb becomes more tender.

As sugar is increased in doughs, another occurrence is alteration of the swelling pattern of wheat starch in dough. It is well known that sucrose is effective in delaying the starch gelatinization process (15), and this delay increases with increasing sucrose. Although this effect is of great importance in making cakes, its importance in bread is unclear. One may speculate that at higher sugar concentrations, delayed starch gelatinization may have at least a small tenderizing effect on the crumb as the dough is baked.

It is generally accepted today that sugar in dough enhances the keeping properties of the bread (16). However, few quantitative data exist, and the available data were obtained several years ago (17).

Sweet Doughs and Danish Sweet Doughs

Sweet doughs and Danish sweet doughs are other types of yeast-leavened bakery foods. Coffee cakes, cinnamon rolls, Danish pastries, and many similar products are made from such doughs. These doughs contain higher levels of sugar than the standard white pan bread, and they also contain shortening, milk, and eggs. Typical formulas for a sweet dough and a Danish sweet dough are shown in Table 9-2.

Table 9-2. Formulas for Sweet Dough and Danish Sweet Dough

Ingredient	Sweet Dough	Danish Sweet Dough
	parts by weight	
Bread flour	100	80
Pastry flour	—	20
Sugar	15	18
Shortening	12	12
Salt	1.5	1.5
Dairy ingredient	2.0	5
Whole eggs	10	12
Water	49	46
Yeast	7	8
Roll-in fat	—	3 oz/lb dough

Procedures

Sweet Dough

Mix all ingredients for 14 to 20 minutes (depending on mixer) at 80°F. Ferment for 30 minutes. Make up into units desired. Pan. Proof for 30 to 45 minutes (104°F, 85% RH). Bake at 400°F.

DANISH SWEET DOUGH

Cream sugar, shortening, salt, and dairy ingredient until light. Add eggs, water, yeast (suspend yeast in water) gradually. Add flour and mix to a smooth dough at 60°F. The dough is folded into a rectangular shape and rested for 10 minutes. Roll-in fat (butter or margarine) is spotted over ⅔ of dough surface, then dough is given book fold. Dough is sheeted, about 0.5 inch thick, three times. Between foldings, dough is best kept in retarder for a rest period of 1 hour or more. After final rest period, dough is made up into desired Danish units, then proofed and baked as for sweet doughs.

The main difference between sweet dough and Danish sweet dough is that the latter is usually made from a richer formula and is laminated. The lamination process involves placing some roll-in fat (usually butter or margarine) on a portion of the sheeted-out dough surface, then folding and refolding the dough in such a way as to ultimately obtain a sheeted-out dough that is comprised of discrete, alternating layers of fat and dough. During baking, thermal expansion of moisture and gases between the layers causes a separation, leading to the flaky character typical of Danish sweet goods. Production of Danish sweet doughs is more labor-intensive than production of regular sweet doughs, but the quality of the former is usually considered much superior.

High sugar levels in these doughs have several consequences. First, the products have a rich, satisfying flavor and a tender, moist crumb. Because of osmotic effects of the higher sugar levels in these doughs, yeast fermentation is slowed down and higher yeast levels must be used to maintain adequate fermentation activity. Also, the higher sugar levels inhibit gluten hydration during the mixing process, so to obtain proper dough development, sweet doughs are usually mixed about 50 percent longer than a typical bread dough.

Chemically Leavened Bakery Foods

This section is concerned primarily with bakery foods that are mostly, but not entirely, chemically leavened and are made with soft wheat flour. (In contrast, breads are made with hard wheat flour.) Only two types of products will be discussed within this broad category, namely cakes and cookies. Both of these are characterized by their rather high levels of sugar.

Use of Sugar in Cakes

Cakes are made with widely varying formulations, methods of preparation and, of course, finished properties. Cakes typically contain high levels of sugar and other ingredients such as eggs, shortening, milk, and various flavoring agents, as well as soft wheat flour. The net result is that cakes have sweet flavors and aromas, and tender textures.

Sugar performs a number of critical functions in cakes, aside from that of flavor. Sugar acts as a creaming and whipping aid in some cake systems, modifies crumb texture and physical properties, aids in formation of crust color by processes similar to those described for bread, and helps moisture retention.

In general, cakes may be classified into two somewhat arbitrary categories, those made with shortening and those made without shortening, that is, foam cakes whose leavening primarily depends on the foaming and aerating properties of eggs.

Shortening Cakes

Table 9-3 summarizes formulas for several shortening-based cakes, including yellow, white, and devil's food layer cakes, and pound cake.

Table 9-3. Formulas for Shortening-Containing Cakes

Ingredient	Yellow Cake	White Cake	Devil's Food Cake	Pound Cake
		parts by weight		
Sugar	130	130	140	120
Flour, cake	100	100	100	100
Shortening, emulsified	45	45	45	60
Whole eggs, liquid	60	—	45	70
Egg whites, liquid	—	67.5	20	—
Nonfat dry milk	10	10	10	5
Baking powder	5	6.25	5	0.6
Soda	—	—	0.63	—
Salt	3.1	3.1	3.75	3.1
Flavor	to suit	to suit	to suit	to suit
Cocoa, dutched	—	—	20	—
Water	80	75	110	45

The sugar levels in these formulas are based on the level of flour and range from 120 to 140 percent. Years ago, cakes were made with less sugar, 100 percent or less based on flour. The introduction of special chlorinated cake flours in the 1920s and 1930s (18) caused great changes in the way cakes were made, permitting the use of significantly higher levels of sugar and liquid. For white or yellow cakes, the upper limit of sugar is roughly 140 percent. The average usage rate is about 120 percent. Cakes made with cocoa or chocolate can take more sugar, perhaps up to 180 percent.

Shortening-based cakes may be divided into three categories, depending upon the method of preparation. Cakes can be made by the creaming method, single-stage mixing, or mechanical mixing.

The creaming method includes sugar directly in the mixing process. In this classic procedure, sugar and fat are first mixed together. Sugar helps to cream air into the fat, which holds the air in the many, very small pockets (cells) developed by the creaming process. These cells are the basis for the subsequent porous nature of the cake structure. After creaming, the remaining liquid and dry ingredients are incorporated into the batter in alternate stages. Batters made by the creaming method are quite stable and produce cakes with fine grain and texture. When the batter goes in the oven, the shortening melts and the air cells migrate into the aqueous phase, where they then can receive carbon dioxide from the chemical leaveners. Thus, the cake becomes leavened.

Single-stage mixing involves mixing all the dry and liquid ingredients at once. The use of special emulsifiers allows the direct incorporation of air into the aqueous phase, where it is stabilized by reduced interfacial tension. Mechanical mixing is utilized by large bakeries. In this process, special high-speed mixing machines quickly incorporate air, in the form of fine bubbles, into the batter essentially by a homogenization action.

In all these cake systems, sugar plays an important role during a critical phase of the baking process. After the cake batter is placed in the oven, the viscosity soon begins to decrease as a function of increasing energy input. At about 185° to 195°F, this process reverses as the flour starch begins to gelatinize and absorb several times its weight in water (19), so that a sharp increase in viscosity results. As the starch gelatinizes, it "sets," and the cake structure is established. At about this point in the sequence of events, maximum carbon dioxide production from the chemical leavener occurs. It is crucial that the setting of the structure and the maximum gas production occur in concert; if the cake structure sets too soon, the gas will diffuse out and be lost, and if the structure sets too late, gas will also be lost. In both cases, cake quality suffers.

A well-known effect of sugar is to delay the onset of starch swelling. Figure 9-3 illustrates how increases in sugar concentration lead to a delay in the onset of starch swelling. Further, sucrose is more effective in this respect than other sweeteners (20, 21). In the batter system under discussion, sugar delays starch gelatinization, controlling the temperature at which the batter goes from a fluid to a solid. The net result is that this setting of structure takes place when the carbon dioxide production is at a maximum. Thus, the gas is held in the air cells of the structure, and a cake with a fine, uniform, porous grain is obtained.

Foam-Type Cakes

Table 9-4 shows formulations for three kinds of foam cakes: angel food, sponge, and snack cakes.

The snack cake formula is basically a sponge cake, but the formula is fairly lean. The egg content is somewhat low to provide adequate leavening and, hence, the formula does contain a relatively high amount of baking powder to supplement leavening activity.

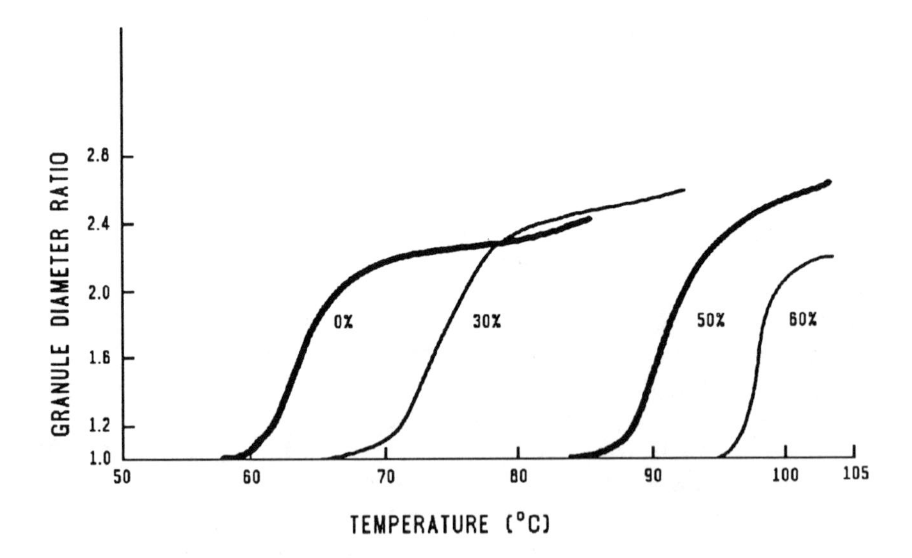

Figure 9-3. Effect of sugar concentration on starch swelling. (15)

This type of formula is utilized by large, commercial, wholesale bakers to produce the hand-held yellow snack cakes that are widely distributed throughout the country.

Table 9-4. Foam-Type Cake Formulas

Ingredient	Angel Food	Sponge Cake	Snack Cake
		parts by weight	
Sugar	273	137	106
Flour, cake	100	100	100
Whole eggs	—	92[a]	10.3[c]
Egg whites	291	—	—
Milk	—	69[b]	10[d]
Baking powder	—	3.6	5
Cream of tartar	4.5	—	—
Salt	4.5	2.1	2.5
Soy flour	—	—	1.7
Flavor	to suit	to suit	to suit
Water	—	—	29

[a] liquid whole eggs [b] liquid whole milk [c] dried whole eggs [d] nonfat dry milk

Sugar is as important in foam cakes as it is for other cakes. In both angel food and sponge cakes, sugar serves as a whipping aid. The first mixing stage of both cakes requires whipping eggs (egg whites for angel food; whole eggs for sponge) and sugar into a light foam that serves as the basic structure of these cakes. After the foam is whipped, the flour is carefully folded in so as not to break the foam. The role of the cream of tartar in the angel food cake is to lower pH and, thus, improve foaming ability.

As in the case of other cakes, sugar in the formula provides flavor and some browning.

Use of Sugar in Cookies

Cookies, like most cakes, are chemically leavened, and have rather high levels of sugar and shortenings—but, on the other hand, they have much lower levels of water. Table 9-5 presents representative formulas for deposit and wire-cut cookies, rotary molded-cookies, and chocolate chip cookies (22).

Table 9-5. Representative Cookie Formulas

Ingredient	Deposit and Wire-Cut Cookies	Rotary-Molded Cookies	Chocolate Chip Cookies
		parts by weight	
Flour	100	100	100
Sugar	30–40	40–50	75
Shortening	30–40	35–40	60
Whole eggs, liquid	7.5–10	—	—
Whole eggs, dried	—	2	7
Nonfat dry milk	2.5–5	2	—
Water	10–20	8	17
Sodium bicarbonate	0.5–1	0.5	0.6
Invert sugar	7.5–25	2	—
Lecithin	0.25–0.5	—	—
Salt	1–1.25	1.5	1.25
Chocolate chip	—	—	55

Cookies usually are categorized by their processing method. *Deposit* cookies are made by forcing dough through a series of nozzles and depositing the formed pieces onto a moving belt. *Wire-cut* cookies involve forcing a dough through dies at the bottom of a hopper, where discs are cut off and drop onto a pan or a moving belt. *Rotary-molded* cookies are produced by forcing a stiff dough into molds embossed on a rotating cylinder; as the cylinder rotates, the molded dough pieces are transferred onto a belt for baking. *Cutting machine* cookies are made from a continuous sheet of dough traveling on a belt; the individual cookies are then cut or stamped from the sheet by reciprocating or rotary cutters.

Cookie doughs for these various processes require differing consistencies for successful production (22). For example, deposit-cookie doughs are relatively soft to facilitate depositing. However, rotary-molded cookie doughs are quite firm so that they will have good molding properties, while still having enough cohesiveness to retain their shape when extracted from the mold.

Sugar is important in cookie formulations for several reasons. First, many cookie doughs are mixed by a creaming process involving sugar and shortening. As for some cake batters, the purpose of creaming is to introduce air into the cookie doughs, and sugar thus serves as a creaming aid. Of course, sugar serves as a flavorant and as a browning agent in cookies, as it does in other bakery foods.

The presence of sugar in the cookie-baking process helps spreading, which has been described by Hoseney (19). When the cookie dough enters the oven, increased temperature causes the dough to become more fluid as the shortening melts. Crystalline sugar in the dough (about half of the formula sugar remains undissolved at the end of mixing) then begins to dissolve with further increases in temperature. As the sugar dissolves, the resultant sugar solution (sugar plus water) increases in volume, and this increase leads to more dough fluidity and cookie spreading. Spreading is further enhanced by the leavening gas that becomes active at this time. Thus, the effects of sugar are critical to spreading, an important quality parameter of cookies (23).

Sugar influences the desirable surface cracking pattern on familiar cookies, such as gingersnaps (19). Figure 9-4 illustrates a cookie of this type. During baking, moisture is rapidly lost from the cookie surface. Sugar will crystallize on the surface and thereby release more water, which also is driven off. As a result, the dried surface cracks as the leavening gases expand the cookies.

Further information on cookies may be obtained from a number of sources (e.g., 22, 24, 25).

Icings and Fillings

Most sweet-type bakery foods are finished by the application of some kind of icing or filling, which has several functions. Icings and fillings enhance the flavor and eating properties of the product; some icings may serve as a barrier to moisture loss from the food; and they embellish the appearance and serve as a merchandizing tool to sell the product.

Of the immense number of icings, glazings, fillings, and related products, only a few can be discussed briefly here. Further information is available elsewhere (6, 22, 26, 27).

Although icings and fillings are composed of a number of ingredients, sugar is the most important. Sugar provides the sweetness and flavor that are the essence of icings and fillings. Further, sugar provides the bulk as well as the structure of most icings and fillings. Boiled icings or fillings may be made with granulated sugar, but those made by a cold process generally require pulverized or powdered sugars, or a fondant sugar, to avoid grittiness and yield a desirable smoothness. Invert sugar in small quantities is utilized at times to provide a gloss or to act as a humectant.

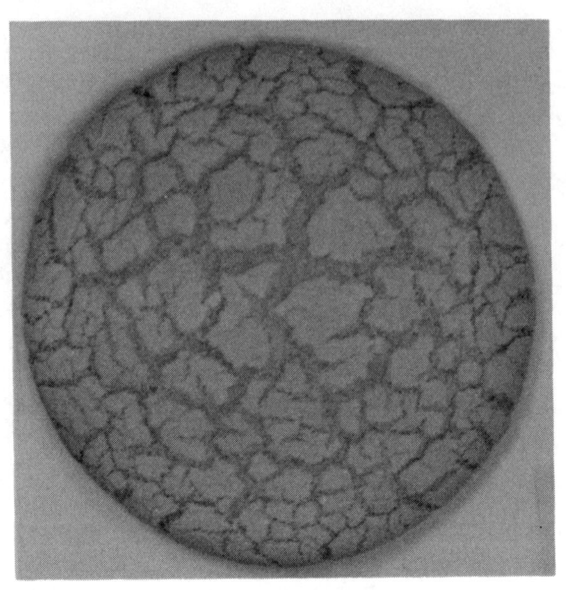

Figure 9-4. Surface cracking pattern of a gingersnap.

Icings

Icings have been defined as modified sugar-water systems in which hydrocolloids and other ingredients are used to control the fine balance between dissolved sugar and suspended sugar and thus ensure a stable system that will not readily break down. Most icings can be categorized as nonaerated (flat), partially aerated, or aerated (foam type).

Table 9-6 shows some representative flat-type icing formulas, recognizing again that a tremendous diversity exists in these formulations. These flat icings are utilized for Danish sweet goods, coffee cakes, and related items. In commercial operations, the icing would be applied by automatic equipment that slowly agitates the icing and keeps it warm to avoid crystallization. Because the final baked product will be sealed in a package, the icing must be stabilized against breakdown during normal shelf-life. The stabilizer indicated in the formula contains gums and starches whose function is to immobilize water in the icing and minimize changes in sugar solubility during product storage. The stabilizer also may contain whiteners, fats, and a filler.

Table 9-6. Flat-Type Icing Formulas

Ingredient	Flat Icing	Vanilla Fudge	Chocolate Fudge
		parts by weight	
Sugar, granulated	25	—	4.5
Sugar, powdered	100	—	10
Sugar, fondant	—	50	—
Sugar, invert	—	2	1
Shortening, emulsified	—	6	1.6
Salad oil	—	—	0.25
Stabilizer	10	—	—
Salt	0.25	0.125	0.1
Water	25	7.75	3.8
Nonfat dry milk	—	2.5	—
Cocoa, dutched	—	—	1.3
Color and flavor	to suit	to suit	to suit

Procedures

FLAT ICING

Boil water and stabilizer for 2 minutes. Add granulated sugar and boil. Add powdered sugar and salt, and mix until smooth.

VANILLA FUDGE

Mix fondant, milk, and water at low speed for 3 minutes. Add the remaining ingredients, and mix at low speed for 7 minutes.

CHOCOLATE FUDGE

Heat granulated sugar and water to 180°F. Add ⅓ of this syrup to remaining ingredients, and mix until smooth. Add rest of syrup slowly.

The formulas for vanilla and chocolate fudge icings are typical of those used on cupcakes and similar items. These icings are not aerated and have a dense, candy-like structure that withstands the abuse of packaging and distribution. The presence of considerable free syrup imparts a smooth, glossy appearance to the icing.

Representative formulas for a partially aerated buttercream-type icing and two aerated marshmallow icings are shown in Table 9-7. Buttercream icings have excellent taste acceptance. They are used on layer cakes and similar items, and also for cake decorating purposes. The shortening used should be designed for fillings and have an appropriate emulsifier system (28). The shortening content of buttercream icings is quite variable and depends on desired attributes. These icings can be supplemented with butter, egg whites, egg yolks and, at times, gums.

Table 9-7. Partially Aerated and Aerated-Type Icings

Ingredient	Buttercream Icing	Marshmallow Icing A	Marshmallow Icing B
		parts by weight	
Sugar, granulated	—	40	40[b] (3.3)[c]
Sugar, powdered	20	—	—
Sugar, invert	—	10	4[b]
Corn syrup, 63 D.E.	—	24	3.3[b]
Shortening, emulsified	10	—	—
Nonfat dry milk	1	—	—
Salt	0.06	0.3	—
Water	3.25	23.4	13.3[b] (1.7)[a]
Flavor and color	to suit	to suit	to suit
Gelatin	—	2	(0.42)[a]
Egg whites	—	—	(13.3)[c]

[a] Stage 1 ingredients [b] Stage 2 ingredients [c] Stage 3 ingredients

Procedure

Buttercream Icing

Place shortening and dry ingredients in bowl, add ⅓ of the water, and mix 5 minutes at low speed. Add balance of water slowly and mix at medium speed to desired aeration.

Marshmallow A

Suspend gelatin in cool water, then heat to 160° to 175°F to dissolve. Blend remaining ingredients and add gelatin solution. Whip until light (specific gravity about 0.25).

Marshmallow B

Stage 1 (gelatin plus 1.7 percent by weight water) dissolved as above. Stage 2 ingredients are heated to 240°F. Stage 3 ingredients are mixed and whipped to a wet peak. Hot Stage 2 solution is added slowly to Stage 3 mixture, which is then whipped until thick. Dissolved gelatin is whipped in.

Marshmallow icings are high in sugar, contain no fat, and are intermediate in moisture content. They are used as coatings for snack cakes, cookies (also as a filling), layer cakes, and other products. Corn syrups are often used in marshmallow icings as crystal modifiers to minimize the growth of sugar crystals. Gelatin, or some other gum, is used to stabilize the highly aerated foam. Marshmallow A icing is made without egg whites and is the type mostly used on snack cakes. Marshmallow B icing utilizes egg whites and is thus more expensive (Table 9-7).

Fillings

Table 9-8 shows two kinds of fillings representative of those widely used by the commercial baking industry.

The cream filling is utilized in various snack cakes that are produced in large quantities by wholesale bakers. Sugar, of course, is the principal ingredient and forms the structure of the filling. Invert sugar is used as a crystal modifier and also to provide a glossy appearance. The shortening is important in achieving proper aeration. The stabilizer contains a combination of starches, gums, and emulsifiers and serves to inhibit migration of the filling into the cake. These fillings are usually fomulated so that the sugar-water ratio provides a water activity of below 0.85, thus ensuring microbiological stability.

The second formula is for sandwich cookie filling. This is relatively simple and is characterized by an absence of water. Cookies have long shelf-lives, and the hard types, such as many of the well-known sandwich cookies, are made with a low-moisture content of about 3 to 4 percent. Thus, the filling cannot contribute moisture to the product. Filling quality can be improved by using higher levels of shortening and perhaps by adding dried milk.

Table 9-8. Filling Formulas

Ingredient	Cream Filling	Sandwich Cookie Filling
	parts by weight	
Sugar, powdered	50	50
Sugar, invert	2	—
Nonfat dry milk	4	0–5
Stabilizer	2.5	—
Shortening, emulsified	18	—
Shortening	—	15–30
Water	8.5	—
Salt	0.125	0.2
Lecithin	—	0.1
Flavor	to suit	to suit

Procedures

CREAM FILLING

Dry ingredients are combined in bowl. Water is added and mixed 6 to 8 minutes until smooth. Add shortening, and mix 10 to 12 minutes at medium speed to a specific gravity of about 0.60 to 0.70.

SANDWICH COOKIE FILLING

Shortening is placed in bowl, mixing is started, and the remaining ingredients are added. Mixing is continued just until a homogeneous mixture is obtained.

Summary

The baking industry is a large consumer of sweeteners. Most bakery foods are made with some type of sweetener, which serves to enhance flavor and other sensory attributes, and performs additional essential functions. Although a number of sweeteners are available for baking purposes, sucrose is the most versatile, in part because it has a high sweetness value, possesses useful physical properties, and can be used in either crystalline or liquid form.

References

1. Pacyniak, B. "Trends—making the right move." *Bakery Production and Marketing*, p. 50. May, 1989.
2. Lahvic, R. "Wholesalers pitch new products." *Bakery Production and Marketing*, p. 34. May, 1988.
3. 1982 Census of Manufacturers. Bakery Products. U.S. Dept. of Commerce, Bureau of the Census. March, 1985.
4. Kulp, K. "Bread industry and processes." In *Wheat: Chemistry and Technology*, 3rd ed. American Association of Cereal Chemists, St. Paul, MN. 1988.
5. Ponte, J.G., Jr., and Reed, G. "Bakery foods." In *Industrial Microbiology*, 4th ed. AVI/Van Nostrand Reinhold, New York. 1982.
6. Matz, S. *Formulas and Processes for Bakers*. Pan-Tech International, McAllen, TX. 1987.
7. Ponte, J. G., Jr. "Production technology of variety breads." In *Variety Breads in the United States*. American Association of Cereal Chemists, St. Paul, MN. 1981.
8. Tang, R. T., Robinson, R. J., and Hurley, W.C. "Quantitative changes in various sugar concentrations during breadmaking." *Baker's Digest* 46(4):48. 1972.
9. Maga, J. A. "Bread flavor." *CRC Critical Reviews Food Technol.* 5:55. 1974.

10. Magoffin, C. D., and Hoseney, R. C. "A review of fermentation." *Baker's Digest* 48(6):22. 1974.
11. Hodge, J. E. "Nonenzymatic browning reactions." In *The Chemistry and Physiology of Flavors*. AVI/Van Nostrand Reinhold, New York. 1967.
12. Namiki, M., and Hayashi, T. "A new mechanism of the Maillard reaction involving sugar fragmentation and free radical formation." In *The Maillard Reaction in Foods and Nutrition*. American Chemical Society, Washington, D.C. 1983.
13. Lane, D. J., and Nursten, H.E. "The variety of odors produced in Maillard systems and how they are influenced by reaction conditions." In *The Maillard Reaction in Foods and Nutrition*. American Chemical Society, Washington, D.C. 1983.
14. Baxter, E. J., and Hester, E. E. "The effect of sucrose on gluten development and the solubility of the proteins of a soft wheat flour." *Cereal Chem.* 35:366. 1958.
15. Bean, M.M., and Yamazaki, W. T. "Wheat starch gelatinization in sugar solution. I. Sucrose: microscopy and viscosity effects." *Cereal Chem.* 55:936. 1978.
16. Kulp, K. "Staling of bread." *Technical Bulletin* I(8). American Institute of Baking, Manhattan, KS. August, 1979.
17. Maga, J.A. "Bread staling." *CRC Critical Reviews Food Technol.* 5:443. 1975.
18. Loving, H. J., and Brenneis, L.J. "Soft wheat uses in the United States." In *Soft Wheat: Production, Breeding, Milling, and Uses*. American Association of Cereal Chemists, St. Paul, MN. 1981.
19. Hoseney, R. C. *Principles of Cereal Science and Technology*. American Association of Cereal Chemists, St. Paul, MN. 1986.
20. Bean, M. M., Yamazaki, W.T., and Donelson, D. H. "Wheat starch gelatinization in sugar solution. II. Fructose, glucose, and sucrose: cake performance." *Cereal Chem.* 55:799. 1978.
21. Spies, R. D., and Hoseney, R. C. "Effects of sugars on starch gelatinization." *Cereal Chem.* 59:129. 1982.
22. Pyler, E. J. *Baking Science and Technology*. Vol. II. Sosland Publishing, Merriam, KS. 1988.
23. Bright, H., Vetter, J. L., and Utt, M. "Effect of sugar and mixing variables on cookie spread." *Technical Bulletin* V(4). American Institute of Baking, Manhattan, KS. April, 1983.
24. Smith, W. H. *Biscuits, Crackers, and Cookies*. Vols. 1 and 2. Applied Science Publishers, Barking, Essex, U.K. 1972.
25. Wade, P. *Biscuits, Cookies, and Crackers*. Vol. 1. Elsevier, New York, NY. 1988.
26. Tressler, D. K. and Sultan, W. J. *Food Products Formulary*. Vol. 2: Cereals, Baked Goods, Dairy and Egg Products. AVI/Van Nostrand Reinhold, New York. 1985.
27. Smith, R. "Update on icings." *Technical Bulletin* XI (3). American Institute of Baking, Manhattan, KS. 1989.
28. Brody, H., and Cochran, W. M. "Shortening for bakery cream icings and cream fillers." *Baker's Digest* 52(6):22. 1978.

10

Sugar in Dairy Products

David E. Smith, Ph.D.*

From a historical perspective, sucrose has been a mainstay in the dairy industry for many years. It is used in a large number of products, anytime added sweetness is desired. These products include various frozen dairy desserts, milks, and yogurt. The form of sucrose, be it liquid, dry, or liquid invert, will vary from one processor to another. Frozen dairy product producers will most likely use either liquid sugar or invert because the large volumes used are much easier handled by pumping and because potential price advantages may be gained from buying in bulk. Processors who make chocolate milk products will probably use the dry form of sugar since the small volumes used do not justify the cost of installation of sugar tanks, lines, and pumps.

Sugar in Frozen Dairy Products

Until the 1940s, sugar was the only sweetener used in large volumes in frozen dairy products such as ice cream. Even today, when other sweeteners may be cheaper than sugar, it is still used widely and is the preferred sweetener in premium ice cream products. The quality attributes that sugar delivers to the product are responsible for its continued widespread use in ice cream products. Sugar helps to produce a clean, sweet taste in ice cream. Corn-derived sweeteners will often produce what is described as a "syrup" taste in the product. Besides imparting a syrupy taste, corn-derived sweeteners may mask or alter the flavorings added to the frozen product.

*David E. Smith, Ph.D., is Associate Professor, Department of Food Science and Nutrition, University of Minnesota, St. Paul, MN.

Most corn-derived sweeteners are not as sweet as sucrose on an equal weight basis and thus cannot be used as a total substitute for sucrose even if flavor is not a consideration. One class of corn-based sweeteners that does have sweetness comparable to sucrose is high fructose corn syrup (HFCS). These syrups, which are usually composed of 42 or 55 percent fructose with the balance of dry solids being mostly glucose and a small percentage of oligosaccharides, have found usage in some frozen dairy products. A major disadvantage of using HFCS, however, is that, on an equal-weight basis, it suppresses the freezing point of the product almost twice as much as sucrose. A lower freezing point will result in less water being frozen at any given temperature, with the result being a product that will develop an icy texture faster.

Ice Cream

Ice cream and related products are processed and produced in the following manner.

1. Ingredients, such as those listed in Tables 10-1 and 10-2, are batched together. The milkfat usually comes from cream or milk with occasional use of butter, butter oil, sweetened condensed milk, condensed milk, and so on. The nonfat milk solids portion of the mix is derived partially from the cream or milk used as the fat source. Also contributing to this part of the product are condensed skim milk, nonfat dry milk, condensed or dried whey, as well as condensed or dried buttermilk. When corn sweeteners are used, they are typically in the 36 to 42 D.E. range. Numerous stabilizers have been used in the past, the more common ones being guar and locust bean gum, carageenan, and various cellulose derivatives. The emulsifiers used include monoglycerides, diglycerides, and polysorbate 80. The liquid ingredients are metered into a tank and, after the temperature has been raised to slightly over 100°F, dry ingredients are blended into the mix.
2. The mix is then pasteurized to kill all pathogenic microorganisms that are present and reduce significantly the total number of microorganisms. The two main pasteurization treatments used are: a) batch pasteurization, which is a minimum of 155°F for 30 minutes, or b) high-temperature, short-time pasteurization, which is a minimum of 175°F for

25 seconds. Recently, some processors have begun to use ultra-high temperature pasteurization treatments to increase the shelf-life of mixes to be distributed to restaurants. The times and the temperatures will vary with shelf-life desired and whether the product is to be refrigerated or not.

3. The mix is then homogenized. Normally a double-stage homogenization is employed to reduce the size of the fat globules. The first stage breaks up large fat globules, while the second stage disperses any clusters of small fat globules that may form following the initial breakup. Pressures used for this treatment will vary depending on product formulation and the equipment being used. The mix is then cooled to 40°F or less.

4. The freezing of the mix involves a reduction in temperature to about 20°F with a simultaneous whipping of sterile air into the product. If liquid flavors are to be added to the product, this is done prior to freezing. Items such as caramel swirls and nuts and berries are added immediately after freezing.

5. The final step is to harden the product at temperatures well below zero to freeze more of the water in the product.

Table 10-1. Examples of Plain Ice Cream Formulas Containing Sucrose as Part of the Sweetening System

Ingredient	Percent by weight[1]				
Milkfat	10.0	10.0	12.0	16.0	18.0
Nonfat milk solids	10.0	10.0	10.0	8.0	7.0
Sugar	16.0	10.0	17.0	17.0	18.0
Corn sweeteners	—	7.0	—	—	—
Stabilizers and/or emulsifiers	0.3	0.3	0.3	0.2	—
Total solids	36.3	37.3	39.3	41.2	43.0

[1]Pounds of ingredients per 100 pounds of product, water making up the balance of the weight.

As previously mentioned, sucrose is an important component of ice cream, especially of the sweetening system. Table 10-1 presents some basic formulas that could be used in the production of ice cream. The first two formulas are typical of economy brand ice creams. Both of these contain the minimum allowable levels of milkfat and nonfat milk solids. The third formula is a more upscale product, while the fourth and fifth formulas represent premium products. In general, the premium products contain more fat, lower amounts of nonfat milk solids, and reduced levels of stabilizers and emulsifiers, if these are used at all. The premium products usually have sucrose as the sole component of the sweetening system because of the clean taste sucrose gives the product as well as a higher freezing point, and thus better textural character. As formulas are cheapened, some of the sucrose is often removed and replaced by corn-derived sweeteners. The total sweetening system usually is in the range of 15 to 18 percent.

Chocolate ice creams have formulas similar to the ones of Table 10-1, the main difference being a reduction in the levels of milkfat and nonfat milk solids. By law, the minimum level of each of these ingredients is 8 percent by weight. Products with reduced milkfat and nonfat milk solids are termed bulky flavors. Cocoa is added to taste, generally in the 2 to 3 percent (by weight) range. Ice creams with added berries and nuts are also considered bulky flavors. Sweetening of these products is accomplished in a manner similar to that used for the plain ice creams. In preparation of such products, one should check the legal requirements regarding level of each ingredient.

Other Frozen Dairy Products

Besides ice cream, there are other frozen dairy-based dessert products. Typical formulas for some of the more popular products are presented in Table 10-2. The two ice milk formulas show the variations that are possible for this product. It should be noted that ice milk can be either hard frozen like ice cream or soft frozen for use in cones. The sherbet formula is a base formula to which flavorings would be added. In addition, citric or lactic acid will be added to produce the tart taste typically associated with this product. The custard formula is similar to those for ice cream with the exception of the added egg yolk solids. Overall, these formulas show a high use of sucrose in the sweetening system. As stated above, sucrose contributes a clean sweetness to the product that does not adversely alter the perception of any added flavoring.

Table 10-2. Examples of Formulas (percent by weight) for Common Frozen Dairy-Based Dessert Products Containing Sucrose as Part of the Sweetening System

	Product			
Ingredient	Ice Milk	Ice Milk	Sherbet[1]	High-Fat Frozen Custard
Milkfat	3.0	6.0	2.0	10.0
Nonfat milk solids	12.0	11.0	3.0	11.0
Sugar	15.0	11.0	18.0	14.0
Corn sweeteners	—	5.0	8.0	—
Stabilizers and/or emulsifiers	0.5	0.5	0.3	0.3
Egg yolk solids	—	—	—	1.4
Total solids	30.5	33.5	31.3	36.7

[1]Titratable acidity of 0.35 is achieved by addition of citric or lactic acids.

One last group of frozen dairy-based products are the specialty products and novelties. Included are frozen yogurts, ices, and ice cream bars and sandwiches. Frozen yogurts are, of course, made from yogurt. One way to make the yogurt is to add 3 to 4 percent nonfat dry milk to whole milk, heat from 180° to 185°F for 30 minutes, cool to 100°F, and add yogurt cultures (setting temperature will vary depending on production time frame). The yogurt is allowed to develop a titratable acidity of about 1.0 percent, expressed as lactic acid by weight, which generally will be accompanied by a pH of 5 or less. Sugar, water, corn syrup solids, and stabilizers are mixed together independently of the yogurt and pasteurized at 165°F for 30 minutes. This mixture and the yogurt are then blended together and homogenized at 2000 and 500 pounds per square inch, first and second stage respectively. The final solids composition of the base to which flavorings are added is given in Table 10-3. Other formulations can be derived by varying the composition of the yogurt, but the sweetener system remains relatively constant. Minor changes can be made to alter sweetness or chewiness of the product.

Ices are produced by blending approximately 12 percent sucrose, 4 percent corn syrup, and 0.3 percent stabilizer in water. Citric acid and flavorings are added before quiescent freezing. Formulations for novelties are as numerous as these products are themselves, and it is hard to give general formulas. The best place to start is an ice cream, ice milk, or frozen yogurt formula; then make modifications until the product meets specifications.

Table 10-3. Solids Composition Frozen Yogurt Base Made From Whole Milk

Ingredient	Percent by weight
Milkfat	2.5
Nonfat Milk solids	9.0
Sugar	17.0
Corn syrup solids (28 D.E.)	8.0
Stabilizer	0.3
	36.8

Chocolate-Flavored Milk Products

Chocolate-flavored milk products are divided into two groups—drinks and milks. A chocolate milk must be made from whole milk, while a drink has milkfat levels as low as 0.5, 1.0, or 2.0 percent. Other than this difference in fat, the processing and other compositional factors are essentially the same for chocolate milks and drinks. There are two basic ways to make these products. The first is to batch the ingredients in their proper proportions prior to homogenization and pasteurization. The second is to make a concentrated slurry of ingredients in some of the milk and then process the slurry. After processing, the slurry is diluted to the proper compositional levels with the rest of the milk. The second method is used when production facilities are not large enough to batch everything at once. There are more problems with settling of cocoa when the second method is used.

The levels of ingredients used in the making of chocolate milk products will vary according to local tastes, but the formula given in Table 10-4 is a good starting point.

Table 10-4. Typical Formula for Chocolate-Flavored Milk Product[1]

Ingredient	Percent by weight
Cocoa	1.0–1.5
Sugar	4.5–6.0
Carageenan	0.025
Vanilla	0.01
Salt	0.02

[1]Ingredients added on w/w basis, with balance of product composed of milk of the desired fat level.

The salt and vanilla, or vanillin, are added to suit local tastes and can be eliminated if desired. Carageenan serves to stabilize the cocoa fibers in solution as well as increase the thickness of the product. Cocoa type will vary from 8 to 23 percent fat and can be either natural or Dutch. Choice of cocoa will again depend on the flavor profile desired in the product. Sugar serves the dual purpose of increasing product viscosity, so as to impart a richer, thicker mouthfeel and modify the flavor of the cocoa and milk to give the characteristic taste of milk chocolate.

Sweetened Condensed Milk

Sweetened condensed milk is cow's milk condensed to a ratio of approximately 2.5:1. The final product contains sugar added to a level of approximately 40 to 45 percent by weight of the finished product. Fat and total milk solids levels are set at not less than 8.5 and 28 percent respectively by federal standards. The following formula can be used to calculate the amount of sugar to add to each batch.

Ratio of sugar to fat in finished product:

$$\% \text{ sugar} + \% \text{ fat} = X \tag{1}$$

$$(\text{Pounds of fat in milk} + \text{pounds of milk}) \times \% \text{ fat} = Y \tag{2}$$

$$X \times Y = \text{pounds of sugar needed} \tag{3}$$

Processing of sweetened condensed milk is done in the following manner:

1. The fat to total solids ratio of the milk is standardized to that desired in the final product. This is usually accomplished by addition of skim milk or cream.
2. The proper amount of sugar as determined above is then added to the product. (Sugar may also be added after condensing. Addition prior to condensing makes the control of boiling in the evaporator easier but may decrease evaporation efficiency.)
3. The product is forewarmed by heating in the range of 180° to 220°F for 5 to 20 minutes. This heating serves not only to pasteurize the product but also to condition the milk proteins to prevent gelling in the finished product.
4. The product is introduced to the evaporator, where it is boiled at reduced pressure until the desired degree of concentration is achieved.
5. After concentration, the product is removed from the evaporator and cooled. This is followed by the addition of a small amount of powdered lactose, which aids in controlling the rate and the extent of lactose crystal growth in the product.
6. Finally, the product is canned and put into storage.

The sucrose in the product helps produce the sweet caramel taste desirable in confectionery products. One other important function of sucrose is its preservative effect on the product. The high level of sucrose increases the osmotic pressure to the extent that no spoilage organisms are able to grow in the properly packaged product.

Egg Nog Products

Egg nog is a seasonal product, produced mostly around the holiday season of Thanksgiving to Christmas. Exact compositional standards will vary from state to state. Some states will make a distinction between egg nog that contains 6.0 to 8.0 percent fat (w/w) and 1.0 percent (w/w) egg yolk solids, and egg nog drink that contains about 3.0 percent fat (w/w) and 0.5 percent (w/w) egg yolk solids. Table 10-5 presents some formulas for egg nog that fit these compositional ranges. It is interesting to note that the overall composition of egg nog is very similar to that of ice milk, except for reduced levels of sweetener.

Table 10-5. Typical Formulations for Egg Nog

Ingredient[1]	Percent by weight		
Milkfat	6.0	8.0	3.0
Nonfat milk solids	13.0	12.0	13.5
Sugar	10.0	12.0	10.0
Corn syrup solids	—	—	4.0
Egg yolk solids	1.0	1.0	0.5
Stabilizer	0.3	0.3	0.3
Total solids	30.3	33.3	31.3

[1]Flavoring can be purchased from variable sources.

Sugar as Part of Flavorings for Dairy Products

In addition to the use of sugar directly in a basic product such as ice cream, chocolate milk, or sweetened condensed milk, sugar will find use in some dairy products as part of an added flavoring system. Two major products that make large use of sugar-containing flavoring are frozen desserts and yogurts. Each of these groups of products will now be discussed in some detail.

Frozen Desserts

Numerous types of sugar-containing flavoring are used in frozen desserts, among which are fruit preparations, caramel and chocolate swirls, candies, cookies, cake bits, various enrobed nuts, etc. A generalized overview of the production of products of this type will be presented. Exact amounts of individual flavorings will vary depending on cost and taste of the target audience. However, some of these products have minimum usage levels defined by law—e.g., nut ice cream must have 2 percent or more nuts on a weight basis, and berry ice creams need to contain 6 percent or more (by weight) berries. Another major concern is the behavior of individual flavorings in the frozen environment of these products. A good example of this is being sure that strawberries are soaked in enough sugar to keep them from freezing too hard. The same can be said of candies and various swirl flavors. There is, however, the possibility of products having too much sugar. This condition creates interference with proper freezing, resulting in a soft, mushy product. The best approach to minimizing these problems of frozen-product hardness is to work closely with the flavor supplier to assure a product tailored for the particular use. Another item of concern, as with all frozen dairy products, is that total sugar, both from the base and from flavorings, is compatible with the amount of acid present. Thus, sensory tests must be conducted to determine the proper acid-to-sweetness balance for each product.

Yogurts

The production of yogurt will be reviewed in more detail before the flavorings for yogurt are discussed. As with any other product, one of the first items of concern is the desired formulation. Currently, there are products in the market that cover a wide range of composition. Some typical formulas for plain yogurts are presented in Table 10-6. Yogurt 1 is a typical skim milk yogurt and would have a weak body. As the content of fat and total solids increases, product body becomes firmer and more custard-like.

Table 10-6. Typical Formulations (percent by weight) for Various Plain Yogurts

Ingredient	Yogurt 1	Yogurt 2	Yogurt 3	Yogurt 4
Milkfat	0.1	2.0	3.5	3.5
Nonfat milk solids	8.8	8.8	8.8	12.0
Total solids	8.9	10.8	12.3	15.5
Water	91.1	89.2	87.7	84.5

Along with the decision on composition of the milk base, the style of yogurt must be decided. Numerous products have been made over the years and will vary from one area of the world to another. In many parts of the world yogurt is of the liquid or drinkable type, while U.S yogurts range from a thick, somewhat flowable product to a firm, gelled product. In addition, it must be decided whether a Sundae-style (fruit on the bottom) or Swiss-style (fruit swirled throughout) yogurt is desired. The rest of this discussion of yogurt will focus on the means of producing Sundae-style and Swiss-style yogurts and will conclude with a brief description of how to modify the flow characteristics of these yogurts.

Figure 10-1 is a flow diagram outlining the various steps for the production of Sundae- and Swiss-style yogurts. Details pertaining to these steps will now be presented and, where appropriate, the differences between these two styles of yogurt will be discussed.

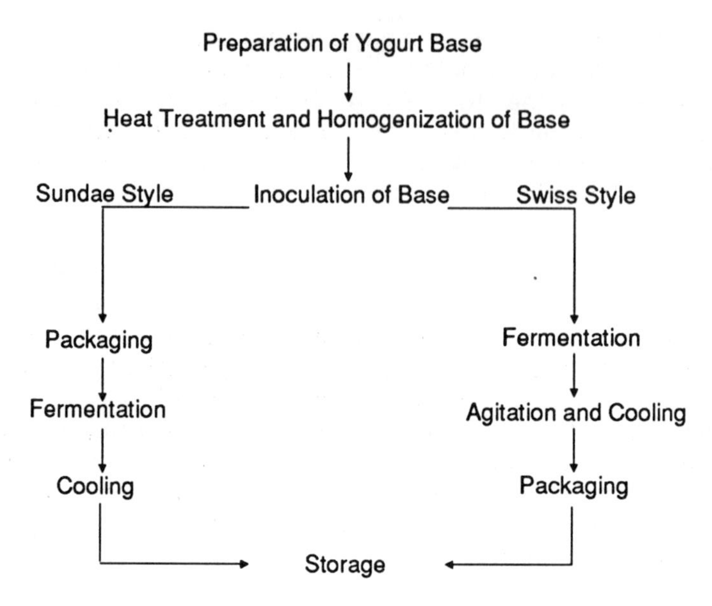

Figure 10-1. Production flow diagram for Sundae- and Swiss-style yogurts.

A number of items go into the preparation of yogurt base. A sweetener would be added at this time if one were desired. In most cases, a plain yogurt would be unsweetened, and flavored products would have sweetener added with the flavoring. If a Swiss-style product is to be produced, a gelling agent such as gelatin needs to be added.

Following the preparation of the base, the product needs to be pasteurized and homogenized. Heat treatment assures that all the pathogenic and nonpathogenic microorganisms have been killed. This treatment is also critical to the denaturing of the proteins in the milk, making them more available to the microorganisms added for fermentation. A typical heat treatment would be 180°F for 30 minutes. Homogenization is done to reduce the size of the fat globules in the milk and thus prevent their rising to the top during fermentation. Typical homogenization pressures used would be 2000 pounds per square inch on the first stage of the homogenizer and 500 pounds per square inch on the second stage.

At this point, the base is ready for inoculation with the culture of starter microorganisms. The temperature is adjusted to 115°F when a fermentation time of four to six hours is desired. If a longer time is desired, the temperature can be lowered slightly. Fermentation temperatures above 115°F can lead to the problem of killing the starter culture. About 2 to 3 percent (w/w) starter is added to the pasteurized and homogenized milk. A typical starter culture is equal amounts of *Lactobacillus bulgaricus* and *Streptococcus thermophilus*. (Activity or concentration should be about 10^9 cells per gram of starter.)

If a Sundae-style product is desired, the inoculated base will be put immediately into the container in which it will be sold. Prior to this, fruit may be placed in the bottom of the container and the base layered on top. Once in the container, the product is placed in a warm room to allow fermentation to take place. When the pH of the product reaches about 4.6, it should have formed a gel and developed an acidic taste. Allowing the pH to drop lower will increase the acidic taste of the product. To stop fermentation, the product is cooled by placing it in a 40°F room. Air flow should be very strong to assure a rapid cooling of the product. Product should then be maintained at this temperature until consumed.

In making a Swiss-style product, the inoculated base is placed in a large tank where the fermentation is allowed to take place. This tank is jacketed, and fermentation temperature is maintained by circulating warm water in the jacket. When the proper pH is reached, the warm water is replaced with cold water, and the agitator on the tank is started. This will rapidly cool the product. The fermented milk is then blended with fruit or other flavoring, filled into the package, and stored at 40°F. The presence of a gelling agent will allow a gel to re-form in the product.

In summary, yogurt production involves the preparation of a base that is pasteurized and homogenized before inoculation with starter culture. These organisms then ferment the product to the desired pH, after which the product is cooled, packaged, and stored until used.

Frozen desserts and yogurts are the two primary dairy-based products utilizing sugar-containing flavoring systems. Other products may have sugar-based flavoring systems. Two such products with limited markets are kefir and fruit-flavored cottage cheese.

Summary

Sweeteners, and in particular sucrose, play an important role in many dairy products. In frozen dairy products, sucrose gives the product a clean, sweet flavor and contributes to its expected creamy body and texture. In addition to sweetness, sucrose is critical to the overall chemical and microbiological quality of sweetened condensed milk. Fluid dairy products rely on sucrose to enhance the cocoa or egg nog flavor added to these products. Thus, it can be said many dairy products would not exist as we know them without sucrose.

References

Arbuckle, W. S. *Ice Cream*, 3rd edition. AVI/Van Nostrand Reinhold, New York. 1977.

Henderson, J. L. *The Fluid-Milk Industry*, 3rd edition. AVI/Van Nostrand Reinhold, New York. 1971.

Hunziker, O. F. *Condensed Milk and Milk Powder*, 7th edition. Published by author. 1949.

Redfern, R. B. and W. S. Arbuckle. *Ice Cream Technology*, 5th edition. Redfern and Associates, Ltd., Raleigh, NC. 1986.

Schmidt, K. A. and D. E. Smith. "Effects of homogenization on sensory characteristics of vanilla ice cream." *J. Dairy Sci.* 71:46. 1988.

Smith, D. E., A. S. Bakshi and C. J. Lomauro. "Changes in freezing point and rheological properties of ice cream mix as a function of sweetener system and whey substitution." *Milchwissenschaft*. 39:455-457. 1984.

Sommer, H. H. *The Theory and Practice of Ice Cream Making*, 5th edition. Published by author. 1947.

Wittinger, S. A. and D. E. Smith. "Effect of sweeteners and stabilizers of selected sensory attributes and shelf-life of ice cream." *J. of Food Sci.* 51:1463-1466. 1986.

Wittinger, S. A. and D. E. Smith. "Effect of sweetener/stabilizer interactions on the viscosity and freezing point of ice cream mix." *Milchwissenschaft*. 41:766. 1986.

11

Sugar in Processed Foods

Nicholas D. Pintauro, Ph.D.*

Sugars are used in processed foods and product formulations, not only for sweetness and as contributors to overall flavor, but also for functional purposes. Sugars impart body and texture, provide bulk, and act as a dispersant or fixative. In addition, there are many complex reactions of sugars with proteins, fats, and other ingredients which can be designed to generate desirable sensory qualities in color and flavor. Sugar (sucrose) is so unique that it is not always possible to use alternative sweeteners. Where substitutions are made, either reformulation or some changes in processing are necessary. Alternative sweeteners include such nonsucrose sweeteners as dextrose (glucose), maltose, corn syrup solids, honey, or molasses.

In processed foods, the functional properties of sucrose and its blends with other sugars can be characterized as follows:

1. soluble in water and readily form crystals upon evaporation;
2. readily fermentable by specific microorganisms;
3. serve as precursors for the development of desirable colors and new flavor entities;
4. serve as dispersants, bulking agents, and dilutants in dry mixes to minimize ingredient separation and reactions between ingredients;
5. serve as agents to decrease water activity (A_w) of dry mixes, thereby improving shelf-life stability of formulated products from chemical and microbiological viewpoints.

*Nicholas D. Pintauro, Ph.D., is Professor Emeritus, Department of Food Sciences, Rutgers – The State University, New Brunswick, NJ.

The technologies described in this chapter will illustrate the functional properties of sucrose in product formulations with emphasis on the use of sugar in its granular or dry form.

A common problem that confronts the food technologist in product development is the selection of the sweetener to use in product formulations such as a fruit-flavored beverage powder, a sweetened tea mix, or a starch-based pie filling. The initial or exploratory step in product development is to use sucrose, which is recognized from experience to be the standard or ideal sweetener. All-sucrose formulations, therefore, are used as targets for comparison with nonsucrose formulations or blends. Partial or whole substitution of sucrose would have to be evaluated in a critical manner for a range of potential problems including:

- solubility and clarity characteristics after dissolving,
- interaction with acids or proteins,
- changes in color due to browning reactions,
- effects on vitamins and mineral fortification,
- changes in product's physical stability (e.g., lumpy or caked condition instead of granular and free-flowing), and
- changes in processing and shelf-life tolerances.

In the last item, "tolerances" refers to the capacity to endure moderate or extreme conditions in processing and storage, especially when prescribed controls for temperature, time, or exposure to high humidity have been exceeded during processing operations or in field distribution.

The problems of caking, vitamin losses, mineral reactions, and browning are well recognized by the food technologist. It is generally known that these problems can be eliminated or minimized if all-sucrose formulations are used. However, there may be economic, or some other, advantages to using sugar blends or a 100 percent nonsucrose system.

Caramelization and Maillard Browning Reactions

Sugars undergo chemical reactions to cause browning and flavor changes known as caramelization. Sucrose is a disaccharide that is hydrolyzable to the two reducing sugars, glucose and fructose. These monosaccharides, at elevated temperatures or prolonged storage, undergo dehydration and internal rearrangement to form compounds called melanoidins. These compounds are readily formed by polymerization to give a wide range of complexes that are brown in color.

The browning reaction occurs also when nitrogen-bearing (N-bearing) materials, such as ammonia salts, proteins (including free amino acids and peptides), and N-bearing vitamins, are present. In both types of browning, the main reaction involves the reducing group (aldehydo or keto) of the sugar. The reaction of the reducing group with amino nitrogen is referred to as the Maillard reaction. This entire series of reactions is also called nonenzymatic browning.

Caramelization and the Maillard reaction take place with or without the presence of acid. The reaction is accelerated, however, in the presence of acid, which acts as a catalyst for the hydrolysis of sucrose. Another important factor for hydrolysis is moisture. Any combination of high moisture, presence of acid, and elevated temperature will cause rapid browning. The browning reaction is time-temperature dependent.

Water absorption by sucrose occurs mainly from the atmosphere during processing and in-package storage, or by transfer from other ingredients in the mix. Sucrose is more stable, in terms of water absorption properties, than dextrose, maltose, and various D.E. grades of corn syrups.

In product mixes containing sucrose, hydrolysis of the nonreducing disaccharide must occur before centers of browning are formed. The use of sucrose instead of corn syrups or dextrose in a product mix such as a beverage or dessert mix provides greater stability or lag time for hydrolysis. Nitrogen or acid impurities become critical if browning reactions are to be avoided when using crystallized sucrose or liquid sugars. Browning reactions can be controlled by several different approaches:

1. Provide critical moisture and temperature control during mixing, processing, and packaging;
2. Provide adequate package protection against moisture penetration of the final product in market distribution;
3. Select less soluble acids and coarser granulations for sugars and N-bearing ingredients;
4. Use coating or encapsulation processes to isolate acids and N-bearing ingredients.

Water Activity and Sugars in Food Manufacture

The availability of water to support the growth of microorganisms in foodstuffs is expressed by measurement of water activity (A_w). The water activity value is defined as that point where the relative humidity (RH) of the atmosphere around the food neither gains nor loses water. At this point of equilibrium the vapor pressure of water in the food is the same as that in the atmosphere.

Water in foods exists in three general forms: (1) free, unbound water; (2) free, immobilized water; (3) chemically bound water. Simple measurements for water content will include free, unbound water and possibly a portion of free, immobilized water. In any consideration of the susceptibility of a food to spoilage or quality losses by microbial action, water activity should be considered in addition to water content.

In addition to the relation of water activity to microbial growth, it can also affect enzyme activity and, in certain instances, chemical activity. In the latter case, critical moisture contents or water activities are associated with such chemical reactions as hydrolysis and the interaction of sugars with acids. Chemical reactions that occur during processing or field distribution should be predicated and analyzed on the basis of both water content and A_w.

A simple expression and definition of A_w (expressed as a decimal) can be given as follows:

$$A_w = \frac{\text{\% RH of atmosphere at equilibrium}}{100}$$

Water activity in a food system can be decreased by substances that dissolve in water. Solutes decrease vapor pressure, so that fewer molecules go into the air above the solution or food. The atmosphere above such a food would have a lower relative humidity than air above pure water or a food with a high vapor pressure (low solids content). Sucrose appreciably decreases the water activity of food systems because its molecular size is relatively small. Dextrose and sodium chloride will reduce water activity even more because of their lower molecular size.

The optimum water activity for the growth of most bacteria and some molds is in the range of 0.995 to 0.990 (pure water has A_w of 1.000). Addition of sugar and/or salt to a moist food will significantly reduce A_w. Jellies and jams have A_w values in the range 0.80 to 0.75 and are susceptible only to mold growth. Self-stable dog food and dry sausage (pepperoni) are other examples of moist foods that are partially preserved because of low A_w formulations containing sugar or salt. Listed in Table 11-1 are typical A_w values for foods containing sucrose or salt systems for food preservation and A_w values for some common dry foods and ingredients.

Table 11-1. Typical A_W Values for Food Systems

Foods with Sucrose or Salt Systems	A_W
40% Sucrose or 7% NaCl	1.00–0.95
55% Sucrose or 12% NaCl	0.95–0.91
65% Sucrose (saturated) or 15% NaCl	0.91–0.87
26% NaCl (saturated)	0.80–0.75
Dry Foods or Dry Ingredients	
Dry noodles (12% water)	0.65
Whole milk powder (2 to 3% water)	less than 0.50
Corn flakes (5% water)	less than 0.50
Granulated sugar (0.5% water)	less than 0.50

Any food with an A_w below 0.50 will not support microbial growth. Also, it has been established, on the basis of experience, that foods or ingredients with an A_w below 0.50 and a water content below 5 percent exhibit stable self-life tolerances. Flavor changes or losses and changes in color or physical condition are nearly nonexistent in foods or ingredients meeting these A_w-moisture criteria.

Agglomeration Processes and Chocolate Drink Products

Agglomeration is a process for the formation of clusters of small individual particles to produce a lacy network. The resultant void spaces between particle clusters allow rapid permeation and absorption of water or other liquids during the solubilization process. An agglomerated dry mix is readily dispersed and dissolved in cold water or milk without clumping. The process of agglomeration also increases particle size without a sacrifice in solubility requirements. It is possible, therefore, to decrease the bulk density of a dry mix by agglomeration; that is, increase the bulk volume of the mix without increasing its bulk weight.

In some instances, dispersibility and solubility can be achieved if a dry mix is added to a liquid with rapid, mechanical stirring. However, with consumer products, it is not advisable to require special equipment or critical recipe handling to disperse a mix in water or milk. Difficulties in achieving complete solubility will often result in product failure and complaints. An example where agglomeration is essential for product performance and acceptability is the addition of cocoa-flavored powder to cold milk for a chocolate milk drink. A typical chocolate drink formulation is given in Table 11-2.

Table 11-2. Prototype Chocolate Drink Formulation

Ingredient (in order of addition)	Pound
Sucrose, fine granular	53
Soya lecithin	1.5
Cocoa, 10–12% fat	12.5
Vanilla flavor (vanilla powder)	1.0
Sucrose, pulverized	26

Sugar is heated and maintained at a temperature above the melting point of the lecithin but below the melting point of sugar. The lecithin is melted and applied in a molten condition to the heated sugar within the range of 130° to 190°F. The mass is stirred to obtain good distribution and coating of the lecithin on the sugar. While lecithin is still in a molten and sticky condition, the cocoa and other ingredients are added while mixing is continued. Finally, the pulverized sugar, which disperses among the clustered granules of the base mix, is added to the stirring mixture. The resulting powder has a greater bulk density than a powder in which lecithin was not used to adhere to the fine granular sugar. This agglomeration process can be accomplished in a steam-jacketed candy kettle equipped with an agitator. Low-pressure steam is used for heating.

Agglomeration of a chocolate drink product can be prepared also by a water-spray process in which the cocoa and sugar particles are wetted sufficiently to adhere to each other. During subsequent mixing (tumbling) and heating, the sugar in the mix is dehydrated and recrystallizes fully in an expanded form. The final moisture content will be 1 to 2 percent. Sucrose and sucrose-invert sugar blends are ideally suited for the water-spray processing of an agglomerated cocoa mix that is readily soluble in cold milk.

Sugars in Fruit Processing

Few fruits are processed (canning and freezing) without the addition of sugar solids. Fresh flavor, aroma, and color are seriously altered without the addition of some type of sugar. In recent years, special processes have been developed for reduced-caloric products, such as dietetic foods, by replacing sugar with an artificial sweetener.

Sucrose and concentrated syrups will withdraw water from tissues of fruit by osmosis. Sugar molecules enter the cellular tissues and form complexes with cell wall polysaccharides. Cell structures are strengthened and protected against weakening or softening as a result of preservation processing, either by thermal or freezing methods.

Enzymatic browning is inhibited by addition of sugar syrup to fruit prior to freezing or canning. Sugar concentrations of 30 to 60 percent are commonly used. More concentrated solutions tend to draw water from the tissues, leading to a shrunken condition. Weaker syrups will be below the osmotic concentration of the fruit juice, and accordingly the fruit tissues will take up water to give a soft texture. The use of ascorbic acid as an antioxidant permits the use of weaker sugar syrups without browning problems.

Sugar syrups can be formed by dissolving dry sugar in the fruit juice itself. In fruit processed by freezing, this is known as dry sugar packing. Syrup packs tend to give better color retention because of immediate protection against exposure to air. Dry sugar packs give a product with a higher drained weight and less liquid after thawing. The usual recommendation for a frozen-fruit dry sugar pack is 3:1; that is, three pounds of fruit to one pound of sugar.

Tomato Catsup, Chili Sauce, and Barbecue Sauce

The U. S. Food and Drug Administration Standards of Identity gives the following definition for tomato catsup: "Tomato catsup is the concentrated product made from the pulp and juice of ripe tomatoes (exclusive of skins, seeds, and cores), vinegar, salt, sugar and/or dextrose, spice and other seasoning."

The process for the preparation of concentrated tomato pulp involves the washing, trimming, and crushing of ripe tomatoes. Two methods, called the "hot break" and the "cold break" method, can be used for crushing. In the hot break method, crushed tomatoes are immediately heated to inactivate enzymes that degrade pectins. Loss of pectins gives a thinner consistency, which is undesirable for catsup-type products. Pectic enzymes are extremely active where even a few minutes' delay in heating would cause significant loss in product consistency. Another factor that controls consistency is the solids content of the end product. Sugar is added to tomato pulp concentrate as a source of flavor and solids in catsup production. Good control of consistency (viscosity) is important so that the product will flow or squeeze properly from the package without appearing thin or watery.

Catsup can be produced from juice or paste. Thick pulp, or juice with about 4 to 5 percent solids, is concentrated to an average of 25 percent solids in a steam-jacketed kettle, vacuum pan, or evaporator. Paste can have 65 percent, or higher, solids. Water is added to dilute the paste to the necessary solids percentage. A prototype tomato catsup formulation is listed in Table 11-3.

Table 11-3. Prototype Tomato Catsup Formulation

Tomato pulp Sp Gravity 1.020, 5% solids (by weight)	800 gal
Sugar	450 lb
Salt	85 lb
Vinegar, 100 grain	36 gal
Oleoresin (spice blend)	to taste
Cook to finish (29–33% solids by weight)	

Care should be taken to dissolve the sugar and salt completely in the tomato pulp. Vinegar and spices are added at the end of the cooking cycle and just before the end of the mixing cycle. During mixing, the product is stirred constantly at 140° to 160°F. The product is filled into standard catsup bottles at 190°F. Sodium benzoate is sometimes added as a preservative.

Barbecue sauce and chili sauce have the same basic formulation as catsup. Adjustments in sugar, vinegar, and in the type and amounts of spices give these products their characteristic aroma and flavor.

Fruit-Flavored Beverage Powders

Beverage mixes can be formulated and processed to be reconstituted without clumping in water by spoon stirring. A beverage ready for consumption is provided in a matter of seconds. The major constituent of such mixes is granulated sugar that has been wetted with a food color solution; for example, orange color is added to sugar for an orange-flavored drink. The color solution is sprayed onto the sugar in the proportion of 500 milliliters or less per 1,000 pounds of sugar. The use of a low amount of color solution eliminates the necessity of a later drying step. In this process, the small amount of surface wetting of the sugar particles is sufficient to give a certain amount of agglomeration, which aids in dispersibility and solubility.

A prototypical vitamin-fortified, orange-flavored dry beverage mix is produced by a process such as the following.

BATCH PROCESS

1,175 lb granulated sugar
  100% through U.S. Standard 20 Screen
  8% max through U.S. Standard 100 Screen
  Moisture less than 0.5%
1 liter FD&C orange color solution
5 lb tricalcium phosphate (TCP)
  100% through U.S. Standard 140 Screen
  5% max through U.S. Standard 200 Screen
  Moisture less than 4%
1 lb vitamin A palmitate
10 lb vitamin C (ascorbic acid)
50 lb clouding agent (dried plasticized fat emulsion in gum arabic)
30 lb flavor emulsion (orange oil in corn syrup)
15 lb trisodium citrate
18 lb sodium carboxymethylcellulose
112 lb citric acid, anhydrous

All ingredients have a particle size approximately equal to that of the granulated sugar.

The coloring solution is sprayed onto the sugar in a ribbon mixer. Mixing is continued for about 10 minutes or until the sugar is uniformly colored. TCP is added next to the tumbling mass. Since TCP adheres to the sugar particles, this ingredient assists in drying the surface of the sugar particles. It is at this point that some agglomeration occurs. The remaining ingredients are added in the order listed in the formula. Mixing is completed in 15 minutes at room temperature and 40 percent relative humidity. The mix has a final moisture content of 0.2 to 0.3 percent. It is free of lumps and free-flowing. The finished product is packaged in glass jars or suitable pouches, sealed for protection against moisture absorption.

Citrus Purée Base for Citrus Beverages

Whole grapefruits are juiced and screened to remove the major portion of tough pulp and rag. The screened juice and fine pulp are homogenized to obtain a purée. A grapefruit purée is prepared according to the formula given in Table 11-4. Final purée charcteristics are also listed.

Table 11-4. Citrus Drink—Prototype Formulation and Analysis

Ingredient/Analysis	
Grapefruit purée	100 parts by weight
Sugar	125 parts by weight
Six-fold lemon concentrate	25 parts by weight
Water	to 100 lb
Final Brix	14.3
Acid	0.75%
pH	2.9

Lemon Pie and Chiffon Fillings

Packaged lemon filling mixes are designed either for institutional or home use. The key features of this line of products are convenience and simple, reliable preparation.

Table 11-5. Lemon and Chiffon Pie Fillings—Prototype Formulation (3½ oz . mix)

Ingredient	Gram
Sugar	59.25
Instant starch	20.00
Carageenan gum	3.00
Fumaric acid	1.50
Salt	1.00
Lemon flavor	To suit
Titanium oxide	0.15
Color	To suit

Instant starch provides smoothness and creamy body. Carageenan gum is the gelling agent. Slight variations in the level of gum can give any desired degree of gel strength or rigidity. Hot water must be used to dissolve the gum. Titanium oxide is used to give the slight milky or opaque appearance characteristic of home preparation recipes. Sweetness can be varied by using blends of sucrose and dextrose.

Instant Creamy "No-Bake" Pie Fillings

A wide range of instant-type pie fillings can be prepared using readily available instant starches. Such starches are soluble in cold water or milk, and thicken and set almost immediately. Creaminess is incorporated into the formulation by the use of a spray-dried topping base. The ingredient line of a typical whip topping base is: "Partially hydrogenated vegetable oil (coconut, palm kernel, soybean), corn syrup solids, sugar, propylene glycol monostearate, sodium caseinate, acetylated monoglycerides, mono- and diglycerides."

Other ingredients in a "no-bake" pie filling formulation are shown in Table 11-6.

Table 11-6. "No-Bake" Creamy Pie Filling—Prototype Formulation

Ingredient	Gram
Sucrose	41.50
Whip topping base (Wip-Treme)	51.00
Nonfat dry milk	8.00
Adipic acid	1.50
Instant starch	2.00
Carboxymethylcellulose (CMC)	0.50
Sodium citrate	2.00
Artificial flavor	0.25
Shade (colorant)	0.03

A typical recipe for a "no-bake" pie filling is: Put the mix (3¾ oz) into a small mixing bowl. Add ½ cup cold milk and mix at low speed for 30 seconds. Beat at high speed for one minute. Add ½ cup cold water and continue to beat for two minutes. Fill into a graham cracker crust shell. Chill. Can be frozen and served in frozen state.

Gelatin Dessert Mix

The ingredient line for a typical gelatin dessert mix reads as follows: "Sugar, gelatin, fumaric acid, sodium citrate, artificial flavor and color." Some brands use a combination of fumaric and adipic acids, while others use adipic acid only. Citric acid is not used for gelatin-based products because of the caking tendencies of citric acid-sugar systems. All brands use sucrose exclusively as the sweetener to avoid browning problems on storage. As explained in a previous section, sucrose provides greater stability than other sweeteners in products that contain a nitrogen-bearing ingredient, which in this case is gelatin.

Table 11-7. Gelatin Dessert—Prototype Formulation

Ingredient	Gram
Sugar	74.65
Gelatin (variable depending on gelatin)	7.00
Fumaric acid	7.75
Sodium citrate	1.00
Artificial flavor (standardized with sugar)	.30
Artificial color (standardized with sugar)	.30
Net weight	85.00

The granulation of the sugar is an important specification for this product. Generally, a granulation specification of no more than 1 percent on a U.S. No. 20 Standard Screen and no more than 60 percent passing through a U.S. No. 100 Standard Screen is required.

The first specification limit ensures that there will be no lumps or coarse particles. The second specification limit controls the amount of fine particles. Granulation specifications are important for several reasons. First, fine granulations tend to absorb moisture more rapidly. This will shorten the shelf-life tolerances for the product. Also, correct particle size distribution is the critical factor in obtaining uniform distribution of the gelatin and other ingredients in the bulk amount of sugar during the mixing and packaging operations.

Lemon-Flavored Iced Tea Mix

Tea extracts are spray-dried to produce an instant tea powder. These powders are soluble in hot or cold water for either hot or iced tea. A popular product is a presweetened instant tea mix with lemon flavor. All that is necessary is to dissolve the powdered mix in water and add ice.

A prototype formulation of a lemon-flavored iced tea mix is given in Table 11-8.

Table 11-8. Lemon-flavored Iced Tea Mix—Prototype Formulation

Ingredient	Percent
Sugar	95.00
Citric acid	2.00
Instant tea	1.75
Caramel color	To suit
Natural lemon flavor	To suit
Tricalcium phosphate	0.50

Citric acid is used for tartness, and in combination with the lemon flavor is a substitute for fresh lemon. Caramel color is coated onto the sugar to give a light brown dry mix which when dissolved gives a rich, deep color like that obtained with a strong brew of tea leaves.

Tricalcium phosphate (TCP) functions primarily as an anticaking agent, and also as a buffer to control pH. Critical adjustment of TCP concentration is sometimes necessary to avoid excessive buffering which will cause flocculation (partial precipitation) of tea solids.

Processed Meats

Sugar and other sweeteners are used in product formulations that include the following classes or categories of cured meats and sausage products:

- Cooked sausage—frankfurters, bologna, salami, pepperoni, loaves
- Cured meats—hams, bacon

The products listed above are processed by curing and require the addition of salt, nitrite, and sugar for the purposes of preservation and flavoring. Spices and smoke flavor are also added for flavor. The sugar concentrations used are not sufficient for preservation. However, sucrose in fermented sausage products functions indirectly as a preservative due to the low pH resulting from fermentation of the sugar to lactic acid. In sausage products, the cure ingredients are added to the meat emulsion prior to stuffing and cooking or smoking. In hams and bacon, the cure is prepared as a solution and pumped into the tissues of the meat before oven or smokehouse processing. Sodium nitrite and nitrate mixtures are used primarily for color control. It has been established that nitrite is more important than salt for preservative action. The preservation factor is important for processed meats. These products are either cooked (equivalent of pasteurization) or semidried for low water activity and, therefore, require refrigeration during storage and distribution.

Sugar is also important in cured meat products to reduce the harshness of the relatively high levels of salt used. High levels of sugar can cause problems in the frying characteristics of bacon and ham due to browning and excessive darkening. Reducing sugars such as dextrose or corn syrups are avoided by some processors because of their increased darkening potential.

Formulations and Processing Methods

Table 11-9. Frankfurter and Bologna Products—Prototype Formulation

Ingredient	lb/100 lb raw emulsion
Boneless beef, 20% fat	65.0
Water (ice)	19.0
Beef trimmings, 50% fat	10.7
Salt	1.7
Sugar	2.0
Spices	1.2
Sodium phosphate	0.4
Sodium nitrite	As prescribed by USDA

Spice mixes or individual seasonings are selected to obtain the distinctive flavors associated with particular brands of frankfurters. The usual spices include pepper, cardamon, sage, garlic, allspice, and clove. Other spices may be used, depending on individual preferences. The meat components are mixed and chopped to give an emulsion. During this operation, the various ingredients listed in the above formulation are added in a prescribed order. Water is added in the form of ice to control temperature, and to obtain the correct particle size and proper emulsion of the fat-protein-water mixture. The emulsion is transferred from the chopper/cutter to a stuffer for pumping into casings for frankfurters or various sized bolognas. The stuffed casings are cooked in a meat processing oven, equivalent to an old-fashioned smokehouse. Artificial smoke may be added to the humidity in the oven or to the raw emulsion. After oven processing, the cooked sausages are spray-cooled and the casing is removed for skinless frankfurters. Texture is an important characteristic for product acceptability. Fat content ranges from 20 to 30 percent and is the most important factor in determining the softness or firmness of texture. The USDA upper limit for fat content is 30 percent.

Dry and Semidry Fermented Sausage—(Salami and Pepperoni)

Formulations for frankfurter-type sausage products do not depend on bacterial fermentation to generate flavor notes. There is a group of popular sausage products which are partially dried (low water activity) and fermented for tangy flavor and an acidic pH. This group includes German- and Italian-style salami (semi-dry) and pepperoni products. Raw batch formulations are similar to those for regular cooked sausages and frankfurters except for increased salt level, different spice blend, and higher pepper level. Special bacterial cultures are added to the chopped meat mixture (beef and pork), which is then stuffed into casings and held for 12 to 48 hours for development of the tangy flavor and acidic pH. A prescribed amount of sugar is added to the raw batch as a nutrient for the culture. Another processing difference for this group of sausage products is the requirement for extended times in special rooms to allow moisture to escape through punctures in the casing. This process permits a gradual lowering in pH and moisture content. Some types of salami do not require any heat processing for shelf-life and food safety requirements. Processing conditions, from raw batch to finished product, require critical step-by-step control to prepare a product of uniform quality and shelf-life.

Boiled and Baked Hams

The so-called "boiled" ham is a bland, nonsmoked product that is used as a luncheon or deli meat. The ham, boneless or bone-in, is pumped with a cure of brine to 110 percent of its uncured (green) weight. The cure is injected by artery pumping or with multiple needles to penetrate the tissues under the fat layer. Other methods are immersion under a cover of brine, or dry-rubbing with the curing salts. The brine solution, or "pickle," contains salt, sugar, and sodium nitrite. Sugar is included at a level of 25 to 50 pounds per 100 gallons of brine.

Individual hams, after boning and curing, are placed in metal containers called retainers or boilers. These containers serve as molds to produce a product of uniform shape and weight. The hams, after sealing in the retainers, are cooked by steaming or immersion in hot water. Hams are cooked at a temperature of 160° to 180°F to achieve a minimum internal ham temperature of 137°F. Some products are cooked as high as 160°F to achieve more flavor and firmness.

Baked ham is the term applied to hams that have been heated in an oven for a sufficient time to assume characteristics of a baked product. As a result, these hams exhibit the formation of a brown surface crust, rendered-out surface fat, and caramelization of sugar applied to the surface. These hams are scored on their fat side, cloves are inserted, and brown sugar is sprinkled dry on the surface.

Bacon

The special cut from the slaughtered hog used for bacon production is called "pork belly." Bellies are pumped with brine by the use of multiple needles inserted into the belly. Sucrose is included in the brine and ranges from 20 to 100 pounds per 100 gallons of brine. Sucrose is added for its flavor and preservative action in combination with salt and nitrite. As with hams, bellies can be cured by immersion or dry-rubbing with the curing mix.

The next step is heat processing and smoking, which takes 12 to 15 hours. Bacon processed by brine pumping yields 98 percent finished product, whereas dry-cured bacon yields 90 to 92 percent because of lower moisture content. In modern production methods, liquid smoke is included in the pump solution and no wood-burning is used in the heat processing step. In place of a smokehouse, artificial smoke and a well-controlled meat processing oven is used. This ensures a more uniform product, less time for heat processing, and reduced shrinkage losses.

References

1. Bennion, M. *The Science of Food.* Harper & Row, New York. 1980.
2. Buckholz, L. "The Role of Maillard Technology in Flavoring Food Products." *Cereal Foods World,* July 1988, American Assn. of Cereal Chemists.
3. Fennema, O. *Principles of Food Science, Part 1: Food Chemistry.* Marcel Dekker, New York. 1976.
4. Pintauro, N. *Agglomeration Processes in Food Manufacture.* Noyes Data Corp., Park Ridge, NJ. 1972.
5. Potter, N. *Food Science,* 4th ed. AVI/Van Nostrand Reinhold, New York. 1986.
6. Wismer-Pederson, J. "Water," in *The Science of Meat and Meat Products.* Eds. Price, J., and Schweigert, B. Food and Nutrition Press, Westport, CT. 1978.

12

Sugar in Ready-to-Eat Breakfast Cereals

C. E. (Chuck) Walker, Ph.D.*

Ready-to-eat breakfast cereals were first made about a century ago by the industry pioneers, Post and Kellogg, in Battle Creek, Michigan. Many of the first ready-to-eat cereals were the progenitors of modern day Wheaties, Post Toasties, Corn Flakes, Grape Nuts, Puffed Rice and Wheat, and Shredded Wheat. Though originally espoused, often by vegetarians, for their nutritional value and digestibility, their convenience and flavor rapidly became the flame, fanned by a wind of innovative advertising, driving them to success. To this day, the ready-to-eat breakfast cereal industry is a curious mixture of real vs. perceived nutritional, convenience, and economic needs.

While the pre-1900 products tended to be austere, and contained only small amounts of sucrose (37), the demonstration at the 1904 St. Louis World's Fair portended a significant change in the products and their promotion. There, to the roar of a remodeled Spanish-American War cannon, Quaker first publicly produced rice "shot from guns." The puffs were coated with caramel and sold to the amazed onlookers. Puffed Rice and Puffed Wheat were initially viewed as products for the confectionery trade, but by 1909, stripped of their sweet coating, not to reappear for half a century, they became nationwide sensations as ready-to-eat breakfast cereals swept the nation (28).

*C. E. Walker, Ph.D., is Professor, Department of Grain Science and Industry, Kansas State University, Manhattan, KS. This chapter has been reprinted with permission of the American Association of Cereal Chemists, Inc.

The first large-scale commercial blendings of something other than cereal grains for ready-to-eat breakfast cereals were apparently two products manufactured by U.S. Mills in Omaha, Nebraska. The first, a health food product of J. M. McGowan, was Uncle Sam cereal. Originating in 1908, it contained flax seed. The second, the first raisin bran, sold today as Skinner's Raisin Bran, was first marketed in 1926 and contained raisins blended with whole hard red winter wheat flakes. A certain Mr. Skinner, the owner of a pasta manufacturing company in Omaha, was traveling on a train with his new bride in the early 1920's. He noticed that she always ordered ready-to-eat cereal flakes for breakfast, and would open a small packet of raisins and sprinkle some over the flakes. He thought it would be a great idea to combine fruit and cereal in the same box, so he contracted with the Uncle Sam Breakfast Food Company to process the cereal for the Skinner Company. In 1964 the rights were purchased by the Uncle Sam Breakfast Food Company, which changed its name to U.S. Mills, Inc., in 1975 (10).

The process required for production of Skinner's Raisin Bran was developed before the pelleting extruder was common, and simply involved cleaning and cooking locally grown hard red winter wheat in large rotating retorts, much as corn grits were cooked for corn flakes. After a period of tempering, the gelatinized kernels were flaked in roller mills and toasted in a gas-fired oven. Immediately prior to packaging, raisins were blended with the flakes. No sugars were added to the raisins, and it was necessary to use a large amount of hand labor to distribute the raisins uniformly.

The early 1960s saw Quaker regain market share with its extruded corn-oat pillow, lacquered with sugar to remain crisp, called Cap'n Crunch. Quaker returned to the presweetened concept with a vengeance, as Cap'n Crunch outsold each of its 50-some competitors.

Another ready-to-eat breakfast cereal destined to evolve into a familiar sugar-frosted form was Shredded Wheat (5). First conceived in 1890 at Nabisco, its success stimulated Dr. Kellogg, visiting from Battle Creek, to redouble his efforts with ready-to-eat cereals.

The Ready-to-Eat Breakfast Cereal Industry Today

Today, ready-to-eat (R.T.E.) breakfast cereals represent big business (19, 21, 36). Retail sales in 1987 were over $4.5 billion, up 10 percent from 1986, and represented an average per capita purchase of eight pounds, or 128 one-ounce servings, in 1987. Consumers show relatively little brand loyalty, as the major companies in the R.T.E. breakfast cereal industry must spend about 10 percent of every sales dollar on advertising, better than double the rate typical for the food industry, to maintain market share for each of the more than one hundred brands on the market. R.T.E. cereals take an average of 110 lineal feet of shelf space in the typical supermarket (1).

The manufacture of breakfast cereals tends to be a "slow technology" industry. Available publications are generally several years old, and often based upon even older publications. The industry is unusually secretive in comparison with the food industry in general. Most current technical publications are limited to patents, work by nutritionists on the finished products, or to general food engineering principles (15, 16, 17, 30, 33, 34, 35). The specifics of formulation and processing are regarded as proprietary. This is no doubt a result of two factors: (a) the unique and comparatively short history of R.T.E. cereals and (b) the relative safety of low moisture foods, not requiring the intensive governmental regulations common in the dairy, meat, canning, and freezing industries.

Market share is relatively stable among the top three major companies and is likely to remain so. Kellogg is the industry leader with about 42 percent, General Mills follows at about 21 percent, and Post (General Foods) is third at 13 percent. The remaining 25 percent is largely divided among three other companies (Quaker, Ralston, and Nabisco) (1). Smaller regional, private label, and health food processors, such as Sunshine and U.S. Mills, account for only a tiny fraction of the total. This status quo is apt to be maintained by the practices of (a) assigning a manufacturer shelf space on the basis of their prior market share, and (b) grouping brands by manufacturer rather than by type. For example, Kellogg's Corn Flakes is not beside Post Toasties, its closest competitor, where it could easily be compared. This practice is probably unique within the food industry.

Breakfast Cereal Manufacturing Processes

There are four major manufacturing processes for R.T.E. cereals: flaking, granulating, shredding, and puffing. A more detailed description of these various processes is available in the recent article by Fast (14).

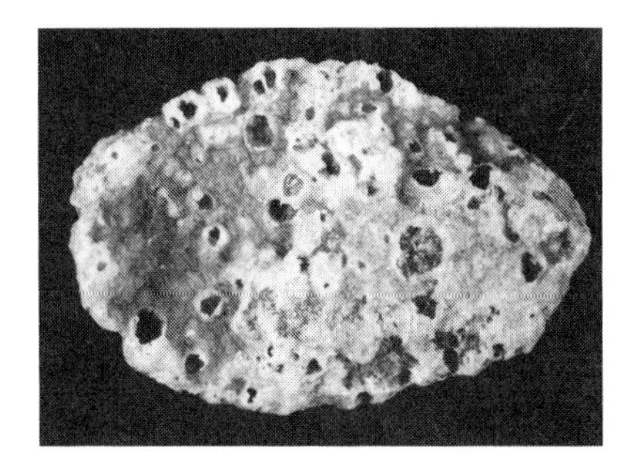

Figure 12-1. White sugar frosting applied to cereal flake. (Kellogg's Frosted Flakes)

Flaking

Flakes probably represent the earliest form of R.T.E. cereals (Figure 12-1). Whole grains may be cooked, flaked between smooth rolls, and toasted; a dough may be formed into pellets, which in turn are flaked and toasted; or a thin gruel may be spattered onto a hot surface, making a crisp flake. The dough-pellet process provides more opportunity for creative formulations, including the incorporation of sucrose, flavors, or binders into the individual cereal pieces, rather than being limited to application of a relatively small amount to the whole grains while they are cooking.

Granulating

Granules and granolas are made by first preparing a stiff dough-like mixture. The dough is then baked or toasted, often followed by some grinding/breaking stages and even an additional toasting or drying. This provides an opportunity to incorporate sugar, either directly into the original dough—for flavor, texture, and color—or for adding it later for a frosting effect. Though it may not be obvious, the increasingly popular granola bar is just a R.T.E. breakfast cereal packaged into a convenient, hand-holdable form (40). The granola bar market exceeded $300 milion in 1987 (23). The division between cereals and snacks appears to be becoming less obvious.

Shredding

Shredding was originally applied to whole grains, especially whole white wheat, which had been cooked until the starch was thoroughly gelatinized. After tempering (holding to allow moisture to equilibrate), the grain was passed between two rolls, one smooth and one grooved, producing a series of endless strings which could be built up into eighteen to twenty layers (Figure 12-2). An alternative

Figure 12-2. Bottom of small shredded wheat biscuit. (Kellogg's Frosted Mini Wheats)

Figure 12-3. Top of small shredded wheat biscuit. This type of manufacture also lends itself to the inclusion of a core of high viscosity jam or fruit filling. (such as Nabisco Strawberry Fruit Wheats)

process involves first mixing a stiff dough, gelatinizing it, and then feeding it through the shredding rolls.

Again, small amounts of sucrose may be cooked with the whole grain, but the greatest opportunity lies in the ability to blend sugar with the other ingredients in the dough stage. Also, a surface sugar frosting may be added before the final drying (Figure 12-3).

A uniquely creative application is the sprinkling of sugar crystals, thick pastes or syrups, and flavorings between the layers of soft strands as the familiar biscuit is built up (Figure 12-4). Sucrose is better than other sugars for layering into shredded products because of its crystalline nature, which retains its identity during drying. The visible crystals contribute a "sparkle" to the product, yet melt quickly when milk is added.

Puffing

Puffing also may be applied to whole grains of wheat and rice, or to pellets made from doughs. General Mills' Kix and Cheerios represent some of the earliest successful cereals of this type. The products may be puffed by modifications, including continuous processes, of the explosive gun puffing process. Another puffing variation includes quickly passing the pregelatinized and slightly flattened (bumped)

Figure 12-4. Interlaced-shredded oat flour biscuit showing sucrose crystals trapped inside during the building-up process. (Quaker Life)

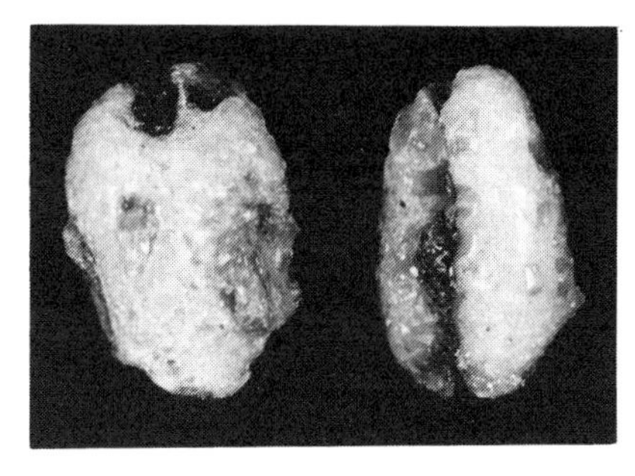

Figure 12-5. Sugar and honey glaze on a puffed whole wheat kernel. (Post Super Golden Crisp)

grains through a very high temperature oven to "oven puff" them. Sugar coatings are often applied to puffed cereals (Figure 12-5).

Extrusion cooking represents the newest technology used by the ready-to-eat breakfast cereal industry. The extrusion process was first used to prepare pellets for subsequent gun puffing, but it was soon followed by the development of cooking-direct expansion extruders. The cooled, directly expanded pieces could be dried easily to below 4 percent moisture, coated with sucrose syrups and/or flavorings, and packaged in an essentially automatic and continuous operation (Figures 12-6, 12-7).

Sugar and R.T.E. Cereals

In a survey of its readers, *Consumer Reports* reported that "taste" led all other reasons for choosing a particular brand of R.T.E. cereal, overshadowing nutrition and far ahead of cost or calories (1). If it tastes good, people will buy it again, and that is one of the main reasons for adding sucrose. Cereal manufacturers make presweetened R.T.E. cereals because that is what people buy. Presweetened products account for approximately 30 percent of the R.T.E. breakfast cereal sales.

Sugar as an Integral Ingredient

Sugar is often added to R.T.E. cereals at the "dough" stage (Tables 12-1 to 12-4). Sugar contributes binding, flavor, and browning characteristics, is critical to controlling the texture and mouthfeel, and acts as a flavor carrier and potentiator of other flavors (18). Total sugar concentration in the dough might run from 6 to 25 percent on a final product dry (3 percent moisture) basis. Blends of sucrose and other sugars are often used, with flavor enhancement the principal contribution of sucrose. While the granulated form of sucrose may be used, a syrup is more easily transported and metered. Maillard browning and caramelization are more likely to occur when the lower-priced reducing sugars are blended with sucrose. This may or may not be desirable, depending upon the product, but it does provide an additional variable for use by the product development scientist (Figure 12-8).

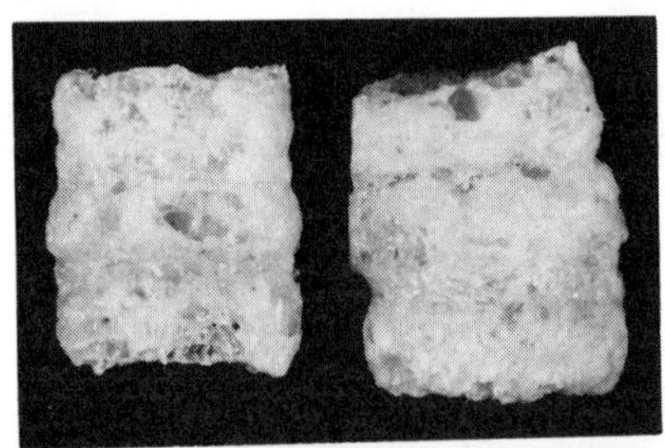

Figure 12-6. Hard sugar glaze on an expanded corn and oat cereal. (Quaker Cap'n Crunch)

Figure 12-7. Expanded oat pellet with a sugar frosting incorporating fruit bits. (Kellogg's Apple Jacks)

Table 12-1. Sample Dough Formula for Corn Flakes (38)

Corn	82.6%
Sucrose	9.5
Rice	4.6
Salt	2.5
Malt	0.8

Water is added to 27 percent moisture. The dough is processed through a continuous extruder to cook and form pellets which can then be partially dried, tempered, flaked, and toasted.

Table 12-2. Sample Dough Formula for Extruded Shapes (2)

Whole wheat flour	44%
Yellow corn cones	44
Sucrose	10
Salt	2

Water is added to 25 percent moisture. The dough is cooked to gelatinize the starch, and pellets are formed at 25 percent moisture content. The pellets may then be flaked and toasted, or puffed by hot air, frying, vacuum, etc.

Table 12-3. Sample Dough Formula for Oat Flakes (8)

Oat flour	60–70%
Rice flour	7–12
Soy flour	5–10
Sucrose	5–15
Lecithin	0.05–0.15
Salt	2–4
Milk protein	1.5–3.5

Water is added to 27–29 percent moisture. Cook at 250°F, 15 pounds per square inch, for 8 minutes. Form dough pellets, dry them to 20 percent moisture content and temper before flaking at 0.01- to 0.05-inch thick. Dry and toast in hot air to 1 to 3.5 percent moisture. Corn flour may be used to replace the oat and rice flour, at 70 to 80 percent, to prepare a form of corn flakes. (Note that a stiff dough of this type may also be shredded.)

Table 12-4. Sample Dough Formula for Hot Air Oven Puffed Corn Pellets (9)

Corn flour	87.6%
Sucrose	2.8
Malt syrup	6.8
Salt	2.8

Water is added to 29 percent moisture. Cook the dough to gelatinize the starch, pelletize, and air dry the pellets to 16 to 21 percent moisture content. Pass the pellets between widely spaced smooth rolls to partly flatten them, and inject them into a 600°F airstream moving at 1000 feet per minute, for 10 seconds, to puff them. Toast at 275° to 300°F for an additional 4 minutes.

Table 12-5. Sample Sugar Coating for a Puffed Wheat Honey Glaze (13)

Sucrose	54.5%
Water	27.2
Invert syrup	7.8
Honey	1.6
Dextrins	8.9

Gradually raise the temperature by evaporation until the boiling point is 325°F. Coat onto puffed wheat to form a glaze, one part glaze to one part wheat.

Table 12-6. Sample Crisp Cereal Coating (31)

Crystalline sucrose	12 oz.
Expanded cereal	10 oz.
7.5% Gelatin spray +	4 oz.
crystalline sucrose	6 oz.

The gelatin and crystalline sucrose mixture is tumbled together and dried to 1 percent moisture. This process uses a noncarbohydrate spray to adhere the sucrose to the expanded cereal pellets, then adds additional crystalline sucrose to the pieces as a "sanding," to keep them from sticking to each other. After drying, the agglomerates are broken up, and the nonadhering sugar is sifted off.

The product remains crisp in the package, yet the sugar dissolves rapidly when milk is added to the serving bowl.

Figure 12-8. Bran cereal using sugars and syrups as a binder, and baked like cookies. (Kellogg's Cracklin' Oat Bran)

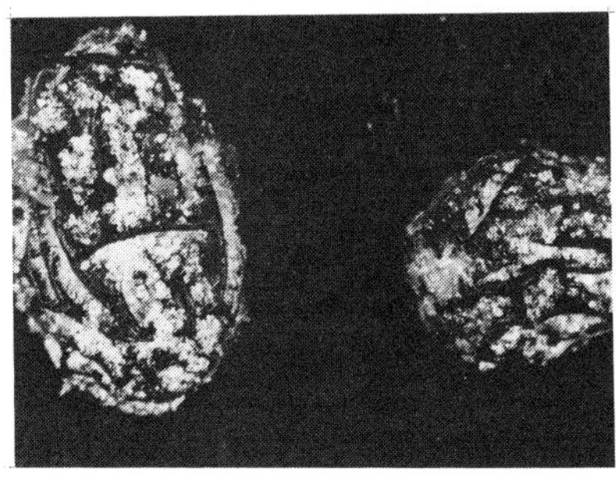

Figure 12-9. Crystalline sucrose applied as a "sanding" on raisins, to prevent clumping or pickup of cereal crumbs. (Kellogg's Raisin Bran)

Sucrose is often also added to grain (corn flaking grits or whole wheat kernels) during cooking, preparatory to flaking. The concentration of sugar is relatively low (about 3 percent). Grains being prepared for oven puffing are similarly treated. Sugar is added principally for flavor enhancement, but may also affect the color, surface porosity, and starch gelatinization.

Sugar in Surface Coatings

By far the major use of sucrose in R.T.E. breakfast cereals is for surface coatings (Table 12-5). The coatings may be in the form of a clear glaze (Figure 12-5), a frosty layer (Figures 12-1, 12-3), or in a powdered or granulated form (Table 12-6). These crystalline/granulated forms also serve as flavor carriers and potentiators, for example with cinnamon or fruit bits (Figure 12-7). Coarsely granulated sucrose is also frequently used to "sand" fruits such as raisins or diced dates, which might otherwise stick together and form nonflowing lumps during packaging or on the shelf (Figure 12-9).

Lawsuits have been filed over whether or not the raisins were sugared, or coated with glycerine and vegetable oil (6).

Sucrose is preferred to other sugars, such as corn syrups, in many surface coatings because of its lower viscosity, fewer problems with excessive browning during final drying, and the ability to crystallize into an attractive "frosting" surface. Blending in small amounts (1 to 8 percent) of viscous corn syrups or honey will control the crystallization rate and grain size, and is used when hard, clear glazes are to be applied (24) (Figures 12-5, 12-6).

Snow-like frostings may be prepared with a syrup of 85 percent sweetener, principally sucrose. These syrups may also contain gelatin or other hydrocolloids. The syrup is atomized onto the toasted cereal prior to final drying under temperature, velocity, and humidity conditions designed to control the rate of crystal growth (42). The crystalline surface may also be controlled by first spraying a syrup solution on the product surface (such as corn flakes), then dusting the surface with a powdered mixture of sucrose and dextrose, and then drying at a temperature low enough to prevent browning (11).

A cereal body with a smooth, nonporous surface will generally require a minimum of 15 percent by weight of added sugar to guarantee a continuous coating. Puffed wheat and rice products, however, are quite porous, and require 25 percent or more by weight to allow for absorption of the added sugar by the cereal body. High contents of invert sugar will cause coated flakes to become sticky, and must be avoided (43).

One unique approach to the sugar coating of cereals is a process using spun (filament) sucrose. The filaments are first made by a process reminiscent of cotton candy (39). The filament fibers are pressed to form a mat, with the cereal sandwiched between two layers, or blankets. When dropped into milk and stirred, the spun sugar dissolves immediately, sweetening the cereal at the same time.

Sucrose Content of R.T.E. Cereals

A number of authors have reported the sugar content of R.T.E. breakfast cereals in the U.S. (26, 27, 41), Australia (20, 22), Canada (25), Denmark (32), Germany (44), and Sweden (3). A recent survey of the sugar content for several cereal types by generic classification has recently been published by the USDA (29) (Table 12-7). These data represent high pressure liquid chromatographic analyses of commercial products purchased at retail.

Total sugar concentrations vary widely, ranging from near 0 to nearly 50 percent. These values are subject to continuous revision, as formulations are changed for economic reasons or in response to market pressures from some who consider simple sugars less desirable than complex carbohydrates. One of the results has been to fragment the ingredient statement by using several different sweeteners, although the total sugar content may not have changed significantly (Table 12-8).

Table 12-7. Total Sugar and Sucrose Content of Some Commercial Ready-to-Eat Breakfast Cereals (derived from Matthews [29])

	% of product as purchased		
Cereal	Total Sugars	Sucrose	% of Sugars as Sucrose
Bran flakes	12.1	9.3	76.9
Bran flakes w/ raisins	26.6	10.1	38
Corn flakes	6.8	2.6	38.2
Corn flakes, sugar coated	39.6	37.6	94.9
Granola w/ raisins	27.4	17.0	62
Oat cereal	2.8	2.8	100
Rice, crispy	8.8	7.6	86.4
Rice, crispy, sugar coated	39.0	37.7	96.7
Rice, puffed	0.1	0.1	100
Wheat & malted barley, flakes	12.4	6.5	52.4
Wheat & malted barley, nuggets	9.1	0	0
Wheat bran	16.4	13.3	81.1
Wheat flakes	9.9	8.2	82.8
Wheat, puffed:			
Plain	1.4	0.6	42.9
Sugar coated	45.1	38.0	84.3
Sugar & honey coated	57.4	44.2	77.0
Wheat, shredded	0.4	0.3	75
Wheat, shredded, frosted	24.6	24.6	100

For more complete listings, including identification of cereals by brand names, see other references cited in (29).

Table 12-8. Presweetened Cereals

Package ingredient statements showing relative precedence of sugar(s) in the final product. (Major ingredients only). Unless otherwise qualified "sugar" = "sucrose." Note that ingredients are listed in decreasing order; for reference, salt will usually be 1 to 2%.

	Cereal Brand Name					
	Cap'n Crunch (Quaker)	Super Golden Crisp (Post)	Frosted Flakes (Kellogg's)	Life (Quaker)	Frosted Mini Wheats (Kellogg's)	Cracklin Oat Bran(Kellogg's)
	corn flour	wheat	corn	oat flour	whole wheat	oat bran/oats
	sugar	sugar	sugar	sugar	sugar	wheat bran
	oat flour	corn syrup	salt	soy flour	sorbitol	brown sugar
	coconut oil	honey	malt flavoring	corn flour	gelatin	partly hydrogenated coconut oil
	brown sugar	partly hydrogenated soy oil	corn syrup	salt		sugar
	salt	salt				corn syrup
						malt flavoring
						coconut
						salt
						baking soda
TOTAL SUGARS	12g/oz=42%	14g/oz=49%	11g/oz=39%	6g/oz=21%	7g/oz=25%	7g/oz=25%

Sugar by the Spoonful

Most people eat sugar on their breakfast cereal. If it is not already present, they add it by the spoonful. An official teaspoon of sugar contains 4 grams. The common table-use teaspoon will hold about 2.5 grams when struck level, but when used to scoop sugar onto cereal it is usually heaping full with as much as 8 grams. One ounce of cereal, even with no sugar present in the cereal, with one heaping teaspoon of sugar will contain 22 percent sugar (i.e., $\frac{8}{28.35 + 8} \times 100$), and about 36 percent sugar when two heaping teaspoons of sugar are added. Children have been observed to use 3 heaping spoonfuls, or 46 percent. Compare these figures with the percent of product as sucrose in Table 12-7. Note that even cereals (such as bran flakes) that are not obviously presweetened may contain about 10 percent sucrose.

It is apparent that sugar added by the consumer can easily exceed (as a percent of total weight basis) the amount added during the production of presweetened R.T.E. cereals.

Conclusion

Sugar is an essential functional ingredient in most R.T.E. breakfast cereals, and contributes an especially significant role in the composition and character of presweetened cereals. Sucrose is the preferred sugar in many surface applications because of its ability to crystallize into a frosty surface, or to form a hard, amorphous, continuous glaze. This extends the potential shelf-life of the cereal, by protecting it from air and moisture, and is popular with many consumers.

References

1. Anon. "Ready to eat cereals." *Consumer Reports* 51:628. 1986.
2. Benson, J.O. "Method of making star-shaped cereal." U.S. Patent 3,077,406. 1963.
3. Birkhed, D., Wange, B. and Edwardsson, S. "Sugars and sugar alcohols in foods." *Var Foeda.* 32:511. (Swedish) 1980.
4. Brody, J.E. *Jane Brody's Nutrition Book.* Bantam Books, New York. 1982.
5. Cahn, W. *Out of the Cracker Barrel: The Nabisco Story.* Simon and Schuster, New York. 1969.

6. Caldwell, E.F. "Raisin bran war settled out of court." *Cereal Foods World* 33(9):807. 1988.
7. Caldwell, E.F. "Good news for cereals." *Cereal Foods World* 34(2):229. 1989.
8. Clausi, A.S., Vollink, W. L. and Michael, E. W. "Breakfast cereal process." U.S. Patent 3,318,705. 1967.
9. Clausi, A.S. and Vollink, W. L. "Preparation of ready-to-eat puffed cereal." U.S. Patent 3,453,115. 1969.
10. Edney, M.J. Personal Communication. U. S. Mills, Inc., Omaha, NE. 1989.
11. Edwards, L.W. "Co-Crystallization of dextrose and sucrose on cereal products." U.S. Patent 4,338,339. 1982.
12. Engstrom, A. and Kern, M. "Breakfast cereals—nutrition and health-related issues." *Cereal Foods World* 25:144. 1980.
13. Fast, R.B. "Process for preparing a coated ready-to-eat cereal product." U.S. Patent 3,318.706. 1967.
14. Fast, R. B. "Breakfast cereals: processed grain for human consumption." *Cereal Foods World* 32:241. 1987.
15. Gehrig, E. J. "The manufacture of cereal flakes-I." *Amer. Miller & Proc.* 92(4):21. 1964.
16. Gehrig, E. J. "The manufacture of cereal flakes-II." *Amer. Miller & Proc.* 92(5):16. 1964.
17. Gehrig, E. J. "The manufacture of cereal flakes-III." *Amer. Miller & Proc.* 92(7):30. 1964.
18. Godshall, M.A. "The multiple roles of carbohydrates in food flavor systems." *Cereal Foods World* 33:913. 1988.
19. Gravani, R.B. "Breakfast cereals today." *Cereal Foods World* 21(10):528. 1976.
20. Greenfield, H., Lee, Y.H., and Wills, R.B.H. "Composition of Australian foods. XI. Mueslis." *Food Tech. in Australia* 33:564. 1981.
21. Hayden, E. B. "Breakfast cereals—trend foods for the 1980's." *Cereal Foods World* 25(4):141. 1980.
22. Jones, G.P., Briggs, D.R., and Toet, H. "The mono- and disaccharide content of some breakfast cereals." *Food Tech. in Australia* 35:281. 1983.
23. Katz, F. R. "In test marketing." *Cereal Foods World* 33:936. 1988.
24. Kent, N.L. *Technology of Cereals.* Pergamon Press, New York. 1983.
25. Korsrud, G. O. and Trick, K.D. "Sucrose, fructose and glucose contents of infant cereals." *Journal of the Canadian Dietetic Association* 40:56. 1979.
26. Li, B. W., and Schuhmann, P.J. "Gas-liquid chromatographic analysis of sugars in ready-to-eat breakfast cereals." *J. Food Sci.* 45:138. 1980.
27. Li, B.W., and Schuhmann, P.J. "Gas chromatographic analysis of sugars in granola cereals." *J. Food Sci.* 46:425. 1981.
28. Marquette, A.F. *Brands, Trademarks and Goodwill: The Story of the Quaker Oats Company.* McGraw-Hill, New York. 1967.

29. Matthews, R.H., Pehrsson, P.R., and Farhat-Sabet, M. "Sugar content of selected foods: individual and total sugars." *Home Economics Research Report #48*. H.N.I.S., U.S.D.A. 1987.
30. McFarlane, I. "In-line measurement and closed-loop control of the color of breakfast cereals." *Cereal Foods World* 33:978. 1988.
31. McKown, W.L. and Zietlow, P.K. "Process for sugar coating ready-to-eat cereal." U.S. Patent 3,615,676. 1971.
32. Meyland, I. "Breakfast cereals. Approximate composition: sugars, dietary fiber and minerals." Publikation, Statens Levnedsmiddel Institut, No. 102. (Danish) 1984.
33. Miller, R.C. "Continuous cooking of breakfast cereals." *Cereal Foods World* 33(3):284. 1988.
34. Miller, B.D.F. "Drying as a unit operation in the processing of ready-to-eat breakfast cereal: I. Basic principles." *Cereal Foods World* 33(3):267. 1988.
35. Miller, B.D.F. "Drying as a unit operation in the processing of breakfast cereals: II. Selecting a Dryer." *Cereal Foods World* 33(3):274. 1988.
36. Poremba, G. "Breakfast cereal market focuses dramatic growth in the eighties." *Processed Prepared Foods* 18(8):36. 1981.
37. Powell, H.B. *The Original Has This Signature—W. K. Kellogg*. Prentice-Hall, New York. 1956.
38. Reinhart, R.D. and Stephenson, R.W. "Method for extrusion cooking of food products." U.S. Patent 3,458,321. 1969.
39. Shoaf, M.D., Groesbeck, C.W., and Cowart, D.G. "Dry food products in spun filaments and method of making same." U.S. Patent 3,615,671. 1971.
40. Sloan, A.E. "Change in breakfast pattern may be among current consumer trends." *Cereal Foods World* 32:246. 1987.
41. Toma, R.B. and Curtis, D.J. "Ready to eat cereals: role in a balanced diet." *Cereal Foods World* 34:(5)387. 1989.
42. Verrico, M.K. "Method and apparatus for spraying snow-like frosting onto food stuff particles." U.S. Patent 4,702,925. 1987.
43. Vollink, W.L. "Process of producing a candy coated cereal." U.S. Patent 2,868,647. 1959.
44. Wolff, J., Nierle, W., El-Baya, A.W., and Fretzdorf, B. "Comparative studies on composition of breakfast cereals." *Getreide, Mehl, und Brot.* (German) 137:13. 1983.
45. Zabek, M.E. "Impact of ready-to-eat cereal consumption on nutrient intake." *Cereal Foods World* 32:234. 1987.

13

Sugar in Beverages

G. J. Marov and J. F. Dowling*

This chapter will discuss sugar in carbonated beverages, powdered drink mixes, and alcoholic beverages.

Carbonated Beverages

During the first three-quarters of the twentieth century, whether in granulated or liquid form, sucrose was the predominant sweetener in soft drinks. Prior to World War II, granulated sugar was the principal sweetener. Later, during the forties, medium invert sugar (MIS), essentially a solution in which one-half the sucrose has been inverted, came to be widely used in the soft drink industry.

Since sucrose was the high-quality original sweetener, it became the standard against which other sweeteners such as corn syrup, dextrose, and later high fructose corn syrup (HFCS) were judged. Corn syrup and dextrose were found to be unsatisfactory not only because of their low sweetness in comparison to sucrose but also because of the modifying effect that these lower-priced sugars had on the flavor of the various beverages in which they were tested. The adverse effects were apparent when these sweeteners were substituted for as little as 20 percent of the sucrose in a beverage formulation. Therefore, until the advent of HFCS, there was no substitute that came even close to displacing sucrose and MIS.

Due to its cost advantage, HFCS replaced sucrose and MIS during the 1980s in most of the soft drinks produced in the United States.

---

*G. J. Marov is a Consultant, formerly Chief Control Chemist and Manager of Quality Assurance for PepsiCo, Inc. J. F. Dowling is Technical Manager, Refined Sugars, Inc., Yonkers, NY.

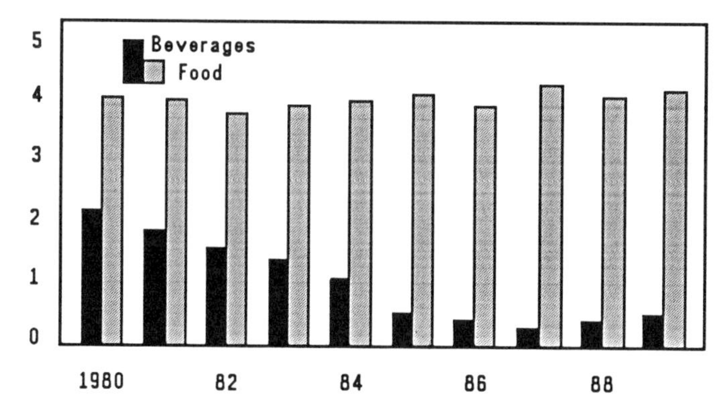

Figure 13-1. U.S. Sugar deliveries for industrial use, 1980-89. The decline in use for beverages is due primarily to increased use of HFCS-55. (Source: U.S. Department of Agriculture, *Sugar and Sweetener Situation and Outlook Report,* Dec. 1989)

(See Figure 13-1.) HFCS-55 is the sweetener of choice in the beverage industry. Sucrose remains the major sweetener in other countries where there is no price advantage for corn. While it has been established that HFCS is a quality sweetener, one should not overlook the fact that sucrose is advantageous in some formulations. New-generation beverage technologists who are starting their careers in the beverage industry should be aware of the physical, chemical, sweetening, and flavoring properties of sucrose and MIS. These properties, as they pertain to beverages, are summarized in this chapter.

Sucrose Formulations

Brix, which by definition is percent sucrose by weight in a pure sucrose-water solution, is the most commonly used term in the soft drink industry. It is the means by which the solids content of sugar syrups, flavored sugar syrups, and beverages is expressed. Since Brix hydrometers and refractometers are calibrated in terms of pure sucrose solutions, Brix measurements by either of these two instruments are an exact measure of the sucrose content of the solution. In this respect, formulations with sucrose are simple: in the case of freshly formulated sucrose syrups and beverages, the measured solids equal the calculated solids. For example, if we prepare a sugar syrup by mixing 3000 grams sucrose with 2500 milliliters (2493 g) water, the calculated Brix, or percent sugar solids by weight, is:

$$\frac{3000}{3000 + 2493} \times 100 = 54.6$$

This is the result obtained, within experimental error, by both the Brix hydrometer and the refractometer.

There is, however, a complication when flavored syrups and beverages are allowed to age at ambient temperature:

1. The measured solids will be higher than expected.
2. Both the hydrometer and the refractometer readings will be lower than the true solids.

These discrepancies occur because of the formation of invert sugar due to sucrose inversion.

Sucrose Inversion

Sucrose differs from all other sweeteners used in soft drinks. It undergoes hydrolysis, which is catalyzed by the acidulant in the flavored syrup and beverage. The reaction is as follows:

$$\underset{\substack{C_{12}H_{22}O_{11} \\ 342}}{\text{Sucrose}} + \underset{\substack{H_2O \\ 18}}{\text{Water}} \xrightarrow[\text{Acid}]{\text{Heat}} \underset{\substack{\text{(Dextrose)} \\ C_6H_{12}O_6 \\ 180}}{\text{Glucose}} + \underset{\substack{\text{(Levulose)} \\ C_6H_{12}O_6 \\ 180}}{\text{Fructose}}$$

The molecular formula and weight of each inversion reactant and product are shown for reference. The above equation indicates that one molecule of sucrose reacts with one molecule of water to yield one molecule each of glucose and fructose, the equal mixture of these sugars being commonly known as invert sugar. This basic sucrose reaction is referred to as "inversion." It is called inversion because, while sucrose is dextrorotatory (it rotates the plane of polarized light to the right), invert sugar is levorotatory (it rotates the plane of polarized to the left). In other words, the optical rotation is inverted from right to left and hence the terminology invert sugar and inversion.

The following factors affect inversion:

1. *pH* The lower the pH, the more rapid the inversion. Therefore, other factors being equal, sucrose in a beverage of pH 2.5 will invert more rapidly than in a beverage of pH 3.0.

2. *Temperature* The higher the temperature, the faster the inversion. Therefore, sucrose in a beverage stored at 90°F for two weeks will be more inverted than in a beverage of the same pH stored at 70°F.
3. *Time* Inversion proceeds with time; therefore, sucrose in an older beverage will be more inverted than in a more recently bottled beverage.

The effects of inversion are:

1. an increase in solids
2. a decrease in volume
3. a gain in sweetness

Increase In Solids. As the above equation indicates, 342 grams of sucrose combine with 18 grams of water to yield 360 grams of invert sugar, so that for every gram of sucrose we obtain $\frac{360}{342}$ or 1.05263 grams of invert sugar. Putting it another way, there is a 5.263 percent gain in solids on total inversion.

The number 1.05263 is the well-known sucrose equivalence factor, which means that 1.05263 pounds of invert sugar are equivalent to 1.00000 pound of sucrose or, conversely, 1 pound of invert sugar is equivalent to $\frac{1}{1.05263}$ or 0.95000 pounds of sucrose.

This factor can be used to calculate the following:

1. The inverted Brix of a flavored syrup or beverage
2. The adjustment in sucrose solids when this sugar is substituted for other sugars in a formulation

The inverted Brix of beverages formulated with sucrose can be calculated by multiplying the beverage Brix by the sucrose equivalence factor. For example:

Fresh beverage Brix:	11.00
Inverted Brix: 11.00 × 1.05263	11.58

Since the nonsucrose components of the beverage do not invert, for accurate results, their contribution to the solids should be subtracted before multiplying by the factor. In the above example, assuming that the nonsugar components contribute 0.3 percent solids, the (inverted Brix) calculation should be made as follows:

$$[(11.00 - 0.3)\ 1.05263] + 0.3 = 11.56$$

The difference between 11.58- and 11.56-Brix is not significant in this particular case. However, in the case of a juice-containing beverage, the error due to the assumption that nonsugar solids are sucrose could be appreciable.

When one sugar is substituted for another in a given formulation, the substitution is sometimes made on a sucrose equivalence basis. For example, assuming that a formulation requires 1000 pounds of HFCS solids or total invert, the following are the equivalent quantities of sucrose and MIS solids that should be used:

$$\text{Sucrose: } 1000 \div 1.05263 = 950.00 \text{ pounds}$$

$$\text{MIS: } 1000 \div 1.02632 = 974.35 \text{ pounds}$$

Although the initial Brix of these beverages will be different (it will be lower for the sucrose formulation), after complete inversion, the solids (inverted Brix) of the beverages will be the same. Of course, sucrose equivalence is not the only basis for substituting one sugar for another. There are other criteria such as adjustments for sweetness, flavor, and mouthfeel differences between sucrose and the substituted sugar.

Decrease in Volume. As the sucrose inversion equation shows, water is used in the reaction. This means that the batch volume will decrease and is generally referred to as "the shrinkage due to inversion." This shrinkage can be calculated as shown in the following example for a 54.00-Brix sucrose solution:

1. *From Sucrose Tables*: 1000 gallons of a 54.00-Brix sucrose solution weigh 10.439 pounds per gallon × 1000 gallons = 10,439 pounds.
2. *On Total Inversion*: The true Brix of the solution becomes $54.00 \times 1.05263 = 56.842$.
3. *From Invert Sugar Tables*: It is found that 56.842-Brix invert sugar weighs 10.543 pounds per gallon.
4. Therefore, dividing the total weight of the solution by its weight per gallon, the volume in gallons of the solution is:

$$10{,}439 \text{ lbs.} \div 10.543 \text{ pounds per gallon} = 990.14 \text{ gallons}$$

5. The shrinkage, therefore, is:

$$1000 - 990.14 = 9.86 \text{ gallons}$$

When this calculation is repeated for sucrose solutions of higher and lower concentrations, the results shown in Table 13-1 are obtained.

Table 13-1. Shrinkage of Sucrose Solutions on Total Inversion

Sucrose Brix	Shrinkage/1000 gal, gal
10.00	2.06
20.00	4.08
40.00	7.83
50.00	9.45
54.00	9.86
60.00	10.66
65.00	11.09
77.00	11.98

The significance of this shrinkage effect is that the contraction in volume will reduce the expected case yield of beverage. It should be noted, however, that the above reductions in volume are maxima; that is, they occur on total inversion of the sucrose. In practice, sucrose is seldom totally inverted because the flavored syrup is usually bottled immediately after preparation or shortly thereafter.

The inversion of sucrose in a flavored syrup as a function of time at 40°F, 70°F, 90°F, and 110°F is given in Figure 13-2. The inversion of sucrose in a beverage as a function of time at 70°F, 85°F, and 100°F is given in Figure 13-3. The refractometer Brix, hydrometer Brix, and true Brix of a flavored syrup and finished beverage are influenced by the level of inversion. These relationships are shown in Figures 13-4 and 13-5, respectively.

As previously mentioned, the solids in flavored syrups and beverages increase with increasing formation of invert sugar. When a hydrometer or refractometer is used to determine Brix, the actual reading will be lower than the true solids (3). Since these instruments are calibrated in terms of sucrose, accurate readings are obtained only in pure sucrose solutions. In syrup solutions that contain invert sugar, both instruments give low readings. In beverages containing invert sugar, low readings are obtained with the

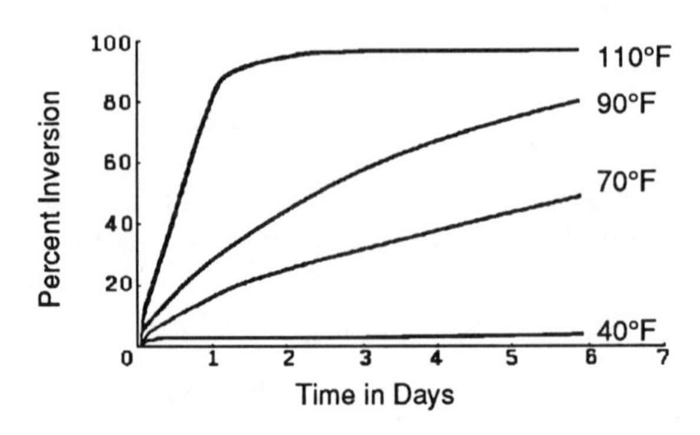

Figure 13-2. Rate of sucrose inversion in flavored syrup. (Based on 1953 laboratory experiments.)

refractometer. It is necessary, therefore, to apply corrections. Figure 13-6 indicates the extent of these corrections. The solid line gives hydrometer corrections while scale-dependent refractometer corrections are shown by the broken and dotted lines. Figure 13-6 gives the following information:

1. As the percentage of solids increases, the difference between hydrometer or refractometer Brix and true solids increases. In other words, more concentrated solutions require larger corrections.
2. Below 10° Brix, the hydrometer reading coincides with the true solids; that is, there is no hydrometer correction. Between 10 and 20 percent solids, the hydrometer correction is negligible (less than 0.05° Brix).
3. It is necessary to apply a refractometer correction even at beverage-level Brix. For example, a correction of +0.1° Brix is required for a 10° Brix beverage.
4. Since the correction for a hydrometer reading is always smaller, hydrometer readings are always nearer to the true solids content than refractometer readings.

It should be noted that Figure 13-6 gives total corrections for pure invert sugar solutions (100 percent invert sugar, on a dry basis). For samples in which the sucrose is only partially inverted, the proper correction may be calculated. For example, the total refractometer

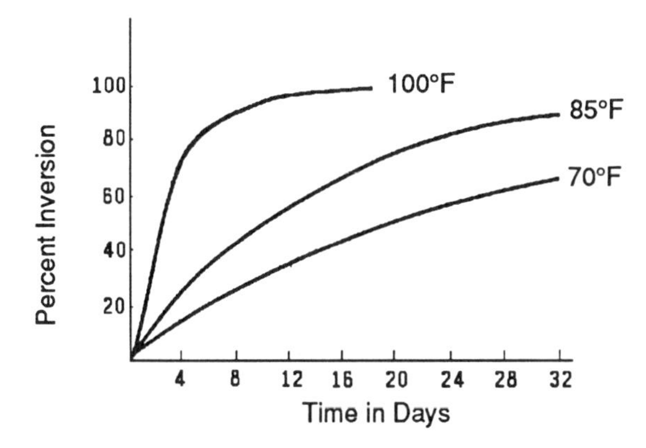

Figure 13-3. Rate of sucrose inversion in beverage. (Based on 1953 laboratory experiments.)

correction at 60° Brix is +1.16; for a sugar solution of this concentration that consists of 80 percent invert sugar and 20 percent sucrose solids, the correction would be 1.16 × 0.80 or +0.93. This correction applies to refractometers calibrated according to the 1936 International Commission for Uniform Methods of Sugar Analysis (ICUMSA) scale. For refractometers calibrated according to the 1974

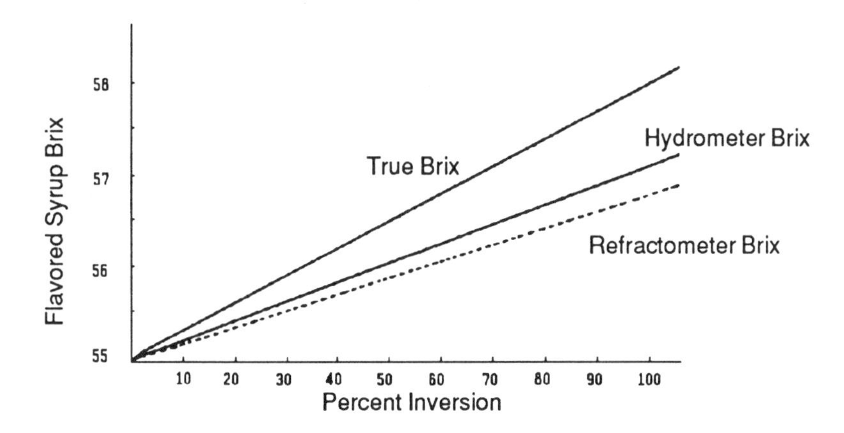

Figure 13-4. Percent inversion vs. Brix in a flavored syrup. (Calculations by G. J. Marov using the 1.05263 sucrose equivalence factor discussed on page 201 in conjunction with data in Table III of Ref. 7.)

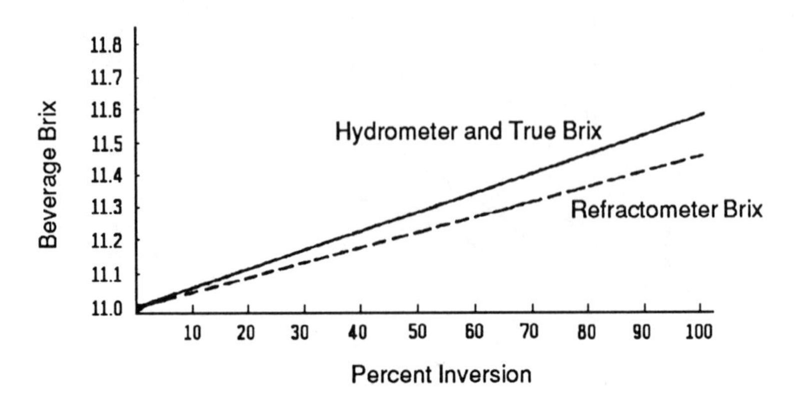

Figure 13-5. Percent inversion vs. Brix in a beverage. (Calculations by G. J. Marov using the 1.05263 sucrose equivalence factor discussed on page 201 in conjunction with data in Table III of Ref. 7.)

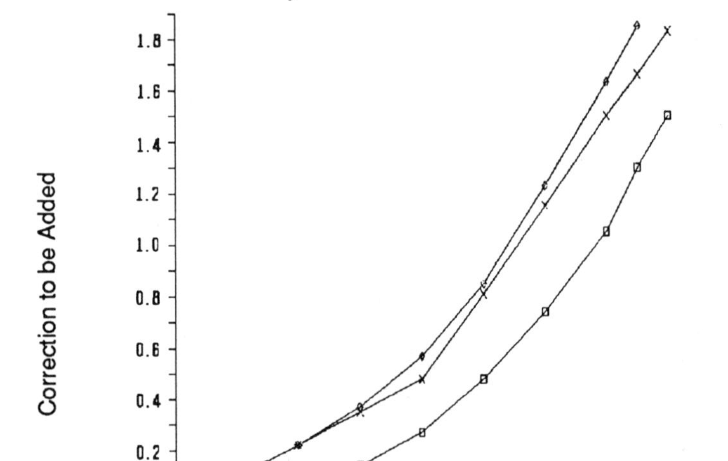

Figure 13-6. Corrections to be added to refractometer or hydrometer Brix of invert sugar to obtain true solids (3).

ICUMSA scale, the total correction at 60° Brix (Figure 13-6) would be +1.23. There are also refractometers in use that are calibrated according to the 1966 ICUMSA scale. However, since the 1966 and 1974 ICUMSA scales do not differ significantly, one correction line is sufficient for both scales (4).

From the above discussion, it is obvious that Brix is a confusing term and may be misinterpreted when used for solutions containing invert sugar. To avoid misunderstandings, the beverage technologist should use terminology such as "refractometer Brix, uncorrected," "Brix, corrected," "percent true solids," or "true Brix."

The beverage technologist should also be aware that the above comments are not only applicable to inverted sucrose solutions. They apply as well to flavored syrups and beverages formulated with HFCS. Hydrometer and refractometer corrections for HFCS are of the same order of magnitude as for invert sugar, but are not exactly the same.

Gain in Sweetness. As sucrose inverts, a carbonated beverage will contain varying amounts of sucrose, glucose, and fructose depending upon the degree of inversion. Sweetness increases until about 50 to 85 percent inversion is reached, after which the sweetness begins to decrease. Although a solution of total invert sugar has 5 percent more solids, its sweetness is equal to that of sucrose (5).

Other Beverage Ingredients

The major ingredient in a carbonated beverage is water, which is close to 90 percent of the beverage. Most flavored, carbonated beverages (cola, ginger ale, lemon-lime) are very delicate flavor systems due to their small quantity of flavoring. Thus, a high-purity water is required to prevent undesirable reactions between water constituents and flavor compounds. The second major ingredient is the sweetener (about 10 percent). All other ingredients present in beverages account for less than 1 percent of the total weight. These ingredients are flavor, acid (citric or phosphoric), color, carbon dioxide, and sometimes a preservative (sodium benzoate). The typical ranges of total solids in selected beverages are shown in Table 13-2.

Bottlers Standards

Due to the delicate nature of carbonated beverage flavors, Bottlers Standards for granulated sugar were developed in 1953 by the American Bottlers of Carbonated Beverages, while standards for liquid sugars were established in 1959. The standards were established to assure that bottlers' criteria for color, percent ash, sediment, and microbiological testing were standardized and uniformly fulfilled. The term "Bottlers Grade Sugar" became standard throughout the industry and refers to sweeteners which comply with bottlers' specifications. Copies of the specifications and methods for testing sugars can be obtained from the sugar companies, or the National Soft Drink Association, 1101 Sixteenth Street NW, Washington, DC 20036.

Table 13-2. Typical Ranges of Total Solids in Selected Beverages and Corresponding Solids per Case (24 8-Oz. Bottles)

Flavor	Solids, % w/w	Solids/Case, lb
Apple	11.0	1.43
Birch Beer	9.8–10.2	1.27–1.33
Celery	9.0–10.0	1.16–1.30
Cherry	11.0–12.0	1.43–1.57
Cherry Cola	11.0–13.0	1.43–1.71
Golden Ginger Ale	9.5–10.0	1.23–1.30
Pale Dry Ginger Ale	8.0–8.5	1.03–1.10
Grape	13.0–14.0	1.71–1.85
Grapefruit	11.6–12.2	1.52–1.60
Cola	9.8–11.0	1.27–1.43
Lemon	11.0	1.43
Lemon-Lime	10.0–11.0	1.23–1.43
Lime	9.0	1.16
Orange	12.6–13.8	1.65–1.82
Pineapple	10.8–12.0	1.41–1.57
Raspberry	11.0–12.0	1.43–1.57
Root Beer	10.0–11.2	1.30–1.46
Sarsaparilla	9.5–10.0	1.23–1.30
Strawberry	12.0	1.57
Tom Collins	9.8–11.0	1.27–1.43

Sugar Flavor

The role of sugar in carbonated beverages is not only to add sweetness but also to contribute body (mouthfeel) and flavor, and to balance the different, dynamic flavor notes so that the full aroma and flavor can be immediately detected by the consumer. A carbonated beverage is a delicate system in which its flavor components are continually changing due to aging and reaction with other beverage components. These changes take place at the same time that sucrose is inverting. Sucrose has the advantage of suppressing certain harsh flavors. Monosaccharides such as glucose and fructose are smaller molecules and will have a different effect on the mellowing and balancing of beverage flavor. Thus, in judging a sugar, one should evaluate it not only as a sweetener but also for its contribution to flavor and flavor stability.

The fact that sucrose contributes more than just sweetness to carbonated beverages became apparent in the 1960s when dietetic beverages were introduced. It was found that simply substituting a nonnutritive sweetener for sucrose did not result in a product that tasted the same as the original beverage. The flavor system had to be reformulated in order to reproduce more closely the taste of the original sucrose-sweetened beverage.

The human being is not a simple scientific instrument and, therefore, taste sensitivity and preference for sweetness vary from individual to individual. Since these measurements are highly subjective, one should test the sugar-formulated beverage in the environment in which it is to be consumed. Since sweetness and flavor relationships change with concentration, time, temperature, acidity, and other factors, beverages should be prepared and tasted over an eight-week period in order to evaluate shelf-life. As discussed previously, sucrose is a flavor enhancer provided that it is not used in high concentrations. As Pancoast and Junk state (6), "Sucrose will serve as an agent for the enhancement of flavor when used in concentrations in which the sense of sweetness will not override the flavors which are being accentuated."

Certain flavor systems, such as ginger ale and cola, are more sensitive to sweetener type than some of the fruit flavors. High-quality beet or cane sugar should have similar flavor profiles. Low-quality sugars (notably those that have high percent ash or color) can lead to flavor and appearance problems during the normal shelf-life of the product.

Powdered Drink Mixes (9)

In recent years, the market for powdered drink mixes has been growing faster than the liquid beverage market due to the convenience of instant tea, dry fruit-flavored mixes, and similar products. Granulated sugar is the leading sweetener for these products because of its low level of nonsucrose materials, heat and microbiological stability, uniform bulk density, controlled grain size, and flavor consistency; furthermore, its uniformly shaped and hard crystals allow proper mixing.

In selecting a sugar grade for dry mixes, a sugar such as X-Fine granulated provides less surface area and lower bulk density than a coarser crystal sugar. A reduced surface area will result in a darker-colored product with the same amount of added color and a more concentrated flavor. Compared to the other granulated sugar products, the grain size of X-Fine granulated gives the optimal surface area for picking up color and flavoring, and dissolves faster in water. In selecting sugar for dry mixes, a uniform sugar grain size distribution is critical. The larger the crystal size distribution, the greater the chance for separation of the dry mix ingredients during handling and packaging.

As previously mentioned, sucrose not only adds sweetness but contributes body and flavor to the drink. The low water activity of the granulated sugar allows for microbiological stability of the dry mix. Bulk granulated sugar can be easily stored, transported, and measured into mixers. Granulated sugar is a high-quality product with an extended shelf-life, and adds stability to the dry mix as a carrier for its other ingredients.

Alcoholic Beverages

Sugar or other carbohydrates can be used as a raw material for the fermentative production of alcohol. In the United States, a portion of the alcohol in beers and fortified wines is generally produced from a cheaper commodity, with corn or rice the material of choice.

Granulated or liquid sugar is used in sweetening wines. Sucrose is preferred because it will add body to the beverage. The liqueur and cordial industries prefer granulated or liquid sugar to add flavor and body to their beverages. In selecting a sweetener for a liqueur, it is important that the sugar be of the highest quality since liqueurs must have an extended shelf-life. A water-white liqueur will become yellow with time if the carbohydrate used does not have the color stability of high-purity granulated sugar. In contrast to sucrose, monosaccharides add less body to the drink and are more prone to browning.

In producing clear liqueurs, the phenomenon of alcohol-induced haze must be recognized. Mixing of a sweetener with alcohol can cause precipitation of residual salts or polysaccharides. Thus, high-purity, low-ash sugar is best for clear, water-white liqueurs.

References

1. Uniform Specific Gravity Table. U. S. Bureau of Standards, Circular No. C457, Dec. 5, 1946. Compiled by Sunkist Growers, Inc., Ontario, CA.
2. Total Invert Table. Supplement to Hoynak, P. X., and Bollenbach, G. N. *This Is Invert Sugar*, 2d ed. CPC International Inc., Englewood Cliffs, NJ. 1966.
3. Marov, G. J. "Hydrometer and Refractometer Brix Corrections for Invert Sugar." Proc. Tenth Annual Meeting, SSDT. 1963.
4. Marov, G. J. "Calibration of Refractometer Brix Scales and Significance of the 1966 and 1974 ICUMSA Tables." Proc. Twenty-ninth Annual Meeting, SSDT. 1982.
5. *Symposium: Sweeteners*. Edited. by G.E. Inglett. AVI/Van Nostrand Reinhold, New York. 1974.
6. Pancoast, H. J., and Junk, W. R. *Handbook of Sugars*, 2d ed., p. 390. AVI/Van Nostrand Reinhold, New York. 1980.
7. Marov, G. J. "A Procedure for Computing and Determining Solids in Syrups and Beverages Formulated with Medium Invert Sugar." *International Sugar J.* 69, 134–136, 163–169. 1967.
8. Woodroof, J. G., and Phillips, G. F. *Beverages: Carbonated and Non-Carbonated*, rev. edition. AVI/Van Nostrand Reinhold, New York. 1981.
9. *Developments in Soft Drinks Technology*, Vol. 3. Edited by H. W. Houghton, Chapter 6, "Powdered Soft Drink Mixes." Elsevier, London and New York. 1983.
10. *Developments in Sweeteners—1*. Edited by C.A.M. Hough, K.J. Parker, and A.J. Vlitos. Chapter 1, "Sucrose—A Royal Carbohydrate." Applied Science Publishers, London. 1979.

14

Sugar in Preserves and Jellies

E. Everett Meschter*

The importation of sucrose into Europe and the development of sugar beet technology made preserving fruits with sugar a viable technology. References to the commercial manufacturing of these products in Portugal and other Mediterranean countries date back hundreds of years. In fact, the word "marmalade" in Portugal originally meant quince preserved in honey (Greek *melimelon—meli* = "honey," *melon* = "apple"), and later sucrose. Today the word commonly refers to preserved citrus fruits, particularly the rind.

Much of what we know about the development of preserves and preserved fruits comes from the discipline of anthropology. Anthropologists have detailed the development of human culture, where people lived, what they ate, and the origin and supply of those things that made up their daily diet.

Sugarcane was grown in Asia at least as early as the fourth century B.C., but evidences of processing (boiling, clarification, and crystallization) date from a millenium later. Sugar that was crudely equivalent to today's product was produced on the southern littoral of the Mediterranean Sea by the eighth century A.D., as well as on the coastal islands and in Spain. It remained costly, was prized less as a food than a medicine, and was perceived much like spices as a flavorant and masker of bitter principles of the herb medicines of the day.

Perhaps the first evidence of the use of sugar as a fruit-preserving agent is referenced in records of a marriage feast in 1403 which included "Perys in Syrippe." "Almost the only way of preserving

*E. Everett Meschter, Consultant to the food industry, Hacienda Heights, CA, was formerly at American Preserve Company, Philadelphia, PA.

fruit," write Drummond and Wilbraham, "was to boil it in syrup and flavor it with spices" (16). Such syrup was made by supersaturating water with sugar and adding spices to taste. Microorganisms that spoil fruit in the absence of sugar can be controlled by a 70 percent sugar solution, which withdraws the water from the cells, killing them by dehydration.

The sugar which became an expensive luxury in medieval times was the product of crushed sugarcane, the same as was known in ancient times. The technology of sugar beets, however, came much later and did not become an important item of commerce until the seventeenth century.

In subsequent centuries, however, the combination of sugar and fruit became more common, and the cost of jams, jellies, fruit spreads, and marmalade declined. Early American cookbooks are replete with recipes for all kinds of fruit preserves, conserves, and fruit butters. In the United States, preserves and fruit butters were made at home and on a cottage industry scale before 1800. Jellies made from extracted juice were a later development.

Although the preservative effect of sugar on the keeping qualities of the finished product is no doubt an important consideration, certainly the added sweetness was instrumental in popularizing these items. Early products contained 35 to 65 percent fruit content and were cooked with sugar to a concentration of 70 to 75 percent soluble solids. The combination of extensive cooking, acidity in the pH range of 3.0 to 4.0, and reduction in water activity resulted in microbiological stability, especially if the products were sealed to exclude air.

With the advent of commercially baked bread and the decline in the time homemakers spent in the kitchen, jam and bread became an important staple in the diets of children and adults alike. In the early 1930s, with a loaf of bread selling for a nickel and a pound of jam selling for a dime, this combination was an important contribution for quick and delicious calories. Jam and bread became the foundation of breakfast toast, lunchtime sandwich, and anytime snack, and even today, it still is.

The depression of the 1930s had an interesting effect upon the jam and jelly industry. This advent of hard times was the motivator of all the ingenuity a company could muster to make products more efficiently. The historical rule of thumb by which preserves were made by the housewife was a cup of fruit to a cup of sugar—the proverbial 45:55 weight:weight ratio that is still prevalent today. As economic pressures became tighter, the unscrupulous packer increased profit margins by removing more of the most expensive ingredient—the fruit. The situation became a threat to the entire industry, and the loss of integrity control was of significant concern to legitimate manufacturers who wanted to see some stability built into meaningful and enforceable "Standards" promulgated by the federal government.

At this point in time, the Bureau of Chemistry, the forerunner of the Food and Drug Administration, addressed the concerns of the preserve industry by analyzing thousands of samples of fruit from all over the country. "Typical" analytical values were established for each fruit variety. These values, in conjunction with analyses of the finished preserves, were used to calculate the pounds of fruit that had actually been used in the production of the product. With the cooperation of the preserve industry, "test packs" were prepared in various plants throughout the country. Formulations were carefully controlled and preserves were made, not only from the 45:55 ratio but also from 35:55 and lower ratios, so that the veracity of the resultant analyses would accurately project the amount of fruit used.

A very interesting summary of this history of the preserve industry is found in the June and July, 1938, issues of *Glass Packer* magazine, edited by Wally Jennsen. These articles describe the methods used by the government to translate chemical results into projections of legitimacy (percentage of fruit) of the formulation.

As recently as 1974, additional changes were made to significantly aid the industry. Since tree fruit products are cooked to 65 percent soluble solids and berry fruits are cooked to 68 percent, there is a significant difference in specific gravity. Jars made to contain two pounds of 68 percent product would overflow when filled with 65 percent product. Conversely, jars made for 65 percent product would be under-filled with 68 percent product. This discrepancy was solved by allowing *all* preserves to be finished to 65 percent solids. Since a berry fruit formulation at 68 percent is "richer" in fruit than at 65 percent, the law was changed to call for the berry category to be made at a 47:55 weight:weight ratio. This resulted in fairness to all interested parties.

Today's Products

Today's products probably do not differ greatly from the preserves and marmalades of centuries ago. Advances in technology and equipment, however, have given rise to better flavor and more pleasing color than were present in the products prepared under more primitive conditions. In fact, these products are remarkably similar worldwide. They all contain around 50 percent by weight of fruit and are cooked to a final soluble solids of 65 to 70 percent.

In addition to the preserves and jellies for which there are U.S. Standards of Identity, other fruit spreads include fruit butters, fruit fillings, and dietary products that deviate from these standards. A product not in compliance with the Standard of Identity must be labelled "Imitation." Therefore, there are many variants of preserves and jellies with fanciful names, such as "Low Calorie Fruit Spread" and "No sugar added" products that use concentrated fruit juice as their sweetening ingredient. Essentially the same caloric value and close replication of the legally defined preserve are achieved when the caloric sweetener is derived from fruit rather than commercial sweeteners. Similarly, there are products sold for dietetic and dietary uses that need not comply with the Federal Standard as long as they are labelled truthfully.

In the United States, the terms preserve and jam are legally synonymous. Traditionally, "jam" is used to denote products made from crushed or comminuted fruit, while "preserves" denote products made from whole or larger pieces of fruit. Jellies are made from the pressed or extracted juice from the whole fruit.

Copies of the Standards of Identity for Preserves, Jams and Fruit Butters (1) are available from the Food and Drug Administration. However, the following are the basic requirements of the Standards.

PRESERVES (JAMS)

47 pounds of berry fruits, 45 pounds of tree fruits.
55 pounds of nutritive sweetener solids (sugar etc.).
Pectin as required to reasonably supplement deficiency.
Food grade acidfying ingredients and buffer salts as necessary to control pH within gelation limits.
Evaporation of water, as necessary, to finish at 65 percent soluble solids (°Brix) by refractometer, temperature corrected.

JELLIES

45 pounds of standard solids fruit juice.
55 pounds of nutritive sweetener solids (sugar etc.).
Pectin as required to reasonably supplement deficiency.
Acidifying agents and buffer salts to control pH within gelation limits.

> Product cooked or diluted, as necessary, to adjust soluble solids to a finished value of 65 percent by refractometer (°Brix). When using concentrated juices, water may need to be added rather than evaporated.

In preserve and jam preparation there are two groups of fruit. The berry fruits (e.g., strawberry, raspberry, blackberry, etc.) require 47 pounds of fruit per 55 pounds of sweetener solids, while tree fruits (e.g., peach, plum, apricot, etc.) require 45 pounds of fruit per 55 pounds of sweetener solids.

Unlike whole fruit, the pounds of fruit juice to be used are based on what the regulations define as a typical solids for that fruit. Grape juice, for example, is considered to be 14.3 percent soluble solids. If the actual Brix is higher than this figure, proportionately less juice may be used. Detailed requirements for all jelly juices are found in the U.S. Food and Drug Administration Jelly Standards.

No added flavors are permitted except such flavor esters that are recovered from the cooking process or from the preparation of fruit juice concentrates from which the jelly is made. Artificial colors and flavors are permitted only in cinnamon- and mint-flavored apple jelly. Preservatives such as sodium benzoate and potassium sorbate are permitted when properly declared on the label.

Although sucrose and honey were the only sweeteners of importance for centuries, modern technology has produced a wide range of sweeteners that contribute unique advantages. The advent of high fructose corn syrup (HFCS) has placed it high on the list of sugar-bearing products used in the preserve and jelly industry. Although HFCS sweetness is comparable to that of sucrose, its major disadvantage is that it is a liquid and may contain as much as 29 percent water (71° Brix), which may have to be evaporated from the final batch. Any boiling or evaporation from a fruit-containing product will of necessity carry a share of the volatile fruit flavors in the distillate. Although products made with fruit juice concentrates will not encounter this problem, HFCS must be considered an important segment of the preserve and jelly sweetener menu.

When sucrose is used as the sole sweetening agent in preserves or jellies, the pH of the system must be in the range of 3.0 to 3.4 to meet pectin setting requirements. At such low pH values the rate of inversion of sucrose to invert sugar is relatively rapid, i.e., completely inverted at room temperature in six months. When the dextrose (glucose) content exceeds 35 percent, there is a strong possibility that it will crystallize in the product and produce an objectionable grittiness. The rate at which inversion takes place must be taken into account to ensure a stable product. Likewise, if sucrose is the sole sweetener and the product concentrated to over 68 percent solids, the solubility of sucrose will have been exceeded and sucrose recrystallization is a distinct possibility.

Following the 1974 Standards amendment, any nutritive sweetener is now permitted, including sucrose, dextrose, corn syrups, and honey. A concentrated fruit juice, such as white grape juice, is not a permitted sweetening ingredient in preserves or jellies.

In the United States, jellies account for nearly 45 percent of the preserve and jelly market. Concord grape is the most popular flavor. Preserves are dominated by strawberry.

The production of commercial spreads totals well over a half-billion pounds, and if various industrial products are included for baked goods, yogurt, and the like, the fruit spread market approximates one billion pounds per year. The usage is largely breakfast-oriented except for sandwiches and peanut butter-and-jelly combinations.

Scientific Principles

While the preparation of preserves and jellies remained a culinary art for many years, the process has been the subject of scientific inquiry. Consequently, commercial production has become more of a science.

Although most fruits contain pectic substances, not all have the quantity or quality required for gelation in the production of preserves and jellies. In addition, if juice concentrates are used, they have probably been completely depectinized prior to concentration and a full complement of the necessary pectin must be added. In the early days of commercial manufacture, the preserver usually added apple or citrus peel, or their extractives, to supplement pectin requirements. Today, high-quality powdered commercial pectins, closely standardized for gel strength, are available and almost universally used.

Pectic substances are a group of heterogeneous polysaccharides with a high molecular weight and a predominant subunit of D-galacturonic acid. These substances are found in all intercellular regions of the cell walls of higher plants. In most plant tissue, and certainly in immature fruit, pectic materials are present in insoluble forms which gradually change to more soluble forms upon ripening.

In fact, the texture of fruits and vegetables is closely related to the solubility of their pectic substances.

The major sources of commercial pectins are apple and citrus peel. After heating to inactivate enzymes, extracting with acid, filtering to yield a clear solution, and concentrating, the partially purified pectin is precipitated with isopropanol, pressed, and shredded. Alternatively, the clear filtrate may be precipitated with aluminum chloride, pressed, and washed with acidified isopropanol. From this point on, the process of pectin production depends upon end product requirements of ester content, amidatation (ammonium hydroxide may be used as a buffer in the process to contribute unique qualities), and classification as a rapid-set, slow-set, or low-methoxyl pectin.

Pectin is a polymer of galacturonic acid whose exposed carboxylic acids are usually methylated to a fairly high degree. Commercial pectins have the ability to increase viscosity or, in sufficient quantity, to form gels. The gelling characteristics of extracted pectins can be modified by removing variable levels of the methyl ester groups, thus exposing more free carboxylic acid groups. Demethylated pectins will form gels in the presence of ions, particularly calcium, and do not require 65 percent soluble solids or any fixed pH environment. Demethylated pectins form gels that are relatively irreversible. Demethylated pectins are largely used for dietary products, although they will gel in a full-sugar product, the texture of which will be softer and quite salve-like. Rapid-set pectins will start to gel at about 180°F while slow-set pectins begin to gel around 160°F.

The studies on pectin prior to 1950 were summarized by Kertesz (4) in a scholarly piece of scientific investigation. Persons interested in this field of study should start with this foundational treatise. More recent work on this subject is covered in *Food Colloids*, with particular emphasis in the chapter on "Commercially Important Pectic Substances," authored by Nelson, Smit, and Wiles (5).

Three factors must be present in order for pectin gelation to occur: pectin, a soluble solids content of 65 percent or higher, and an environment of acidity that gives a pH between 2.2 and 3.8, preferably in the 3.0 to 3.5 range. Although gelation is still possible outside this range, a lower pH may cause gelation in a matter of seconds, while a higher pH may require as much as a month for gelation to occur.

The adjustment of acidity within the pH range required for gelation is dictated by the particular pectin structure. It is pH rather than total acidity that is important in gel formation. The pectins found in fruit and extracted as commercial products require a minimum soluble solids content of 65 percent or, conversely, a water content of 35 percent or less, for gelation. Among the numerous theories on the mechanism of formation of a pectin gel are hydrogen bonding and the dehydrating influence of sugar solids.

The complex association among sugar solids, acidity (pH value), and the chemistry of the commercial pectin is the basis for production of a gelled product of predictable elasticity and strength. A variety of sugars can have a marked effect upon finished gel quality. Sucrose is uniquely uniform the world around, and gels produced with a sugar-based formulation can be duplicated anywhere in the world. Although corn syrups may add unique processing variations, the final gels may possess differing physical characteristics.

Relatively small changes in the amount and type of pectin, pH, and type of soluble solids can have a significant effect on many of the quality attributes of the finished product. If gelling begins before the batch is poured into its final container, the texture will be rough and grainy. This condition is called pre-set and is not reversed by subsequent heating. Further insight into this complex subject is beyond the scope of this chapter, but the bibliography will lead the student to any of the technical phases of pectin gels that may be of interest.

It is obvious that if all these scientific criteria are required to make jellies and jams, our predecessors must have had a great deal of intuition to have been successful. Although these criteria were not completely understood, experience showed that, under certain conditions, adding lemon juice brought success, while other times, more sugar was required to resolve a problem. Needed facts were learned by trial and error.

Quality Attributes

As with many food products, the quality of preserves and jellies can be highly subjective. Individual likes and dislikes as to color, flavor, and consistency can be born from individual experience rather than any objective measures. There are still very few objective tools to adequately evaluate product quality.

The USDA has published Grade Standards for preserves and jellies. Preserves are graded for flavor, color, absence of defects, and consistency. Jellies are graded for flavor, color, and consistency, with flavor rating twice the value of the other criteria. Problems still remain. These observations are human opinion and are therefore not based on objective evaluation. They are, however, widely used by industry and purchasing institutions.

No one expects a strawberry preserve to taste like a fresh strawberry. They are separate and distinct entities. However, the consumer does expect a reasonable characteristic flavor that is free of off-flavors or scorched notes. Similarly, color should be bright, no extraneous leaves or stems should be present, and consistency should be appropriate for the intended use.

The most important reason for quality evaluation is to keep track of product uniformity. A manufacturer would like to have every jar come off the line with perfect scores for all criteria. Since raw material quality varies with nature, season, growing area, and variety, perfection is to be sought but probably never attained.

There are, however, some guidelines that make a high degree of uniformity possible:

1. Use the best-quality raw materials available;
2. Pay attention to the proper preparation of the fruit. Remove defects. Store properly at low temperatures to prevent enzymic deterioration. Juice should be filtered and polished to sparkling clear;
3. Try to use the same variety throughout the year;
4. Control the manufacturing parameters by accurately measuring pH, solids, temperatures, pectin content, etc.;
5. Use a preparation procedure that minimizes the deteriorating effects of times and temperatures;
6. Select quality nonfruit ingredients. For example, the sweetener used should have a positive influence on the texture and quality of the finished product. Sucrose has the clean, clear, bright characteristics to make the finest product.

As noted earlier, the interactions between pectin, acid, and sugar are complex. The use of sucrose yields a product of sharper texture, greater strength, and cleaner flavor than might be expected from the use of higher molecular weight carbohydrates. Furthermore, the virtual absence of reactive monosaccharides (especially fructose) in a sucrose system seems to result in better color and gloss protection. Unfortunately, there is very little published research on these effects.

Modern Manufacturing

Millions of pounds of preserves and jellies have been made in open kettles where the finishing point was determined by viscosity. The old-time processes usually depended on the difference in the boiling characteristics as the product became more concentrated toward the finishing point.

Large-scale processors utilize state-of-the-art technology even though some specialty items are still done by kitchen methods. Ensuring uniformity from batch to batch, and matching or exceeding the quality of the best kitchen production, requires the best tools and procedures that science can provide. Processors now use the refractometer as the routine measure of the end-point. Even in the vacuum pan, where most product is now prepared for protection against color and heat damage, in-line refractometers are built into the side of the pan so that continuous readings can be made and automatically terminate the cooking when the end-point has been reached. Even in-line pH meters can give a continuous and accurate status of the pH of the batch, to control precisely the acidity for uniform gelation.

Most commercial products are now made in stainless steel vacuum equipment. At some point, either by preconcentrating the fruit material or during the actual preparation, water must be removed either by evaporation or, in the case of some juices, by freeze drying. Vacuum process vessels allow the temperature of evaporation to be as low as 120°F. Heat-induced flavor changes and color degradation are minimized.

Preserves and jellies at 65 percent solids have a water activity that is borderline for the prevention of yeast and mold growth. Substitution of high molecular weight carbohydrates for any portion of the sugar renders microbiological stability doubtful. As a result, commercial products are either filled hot into jars which are then hermetically sealed, or filled at a moderate temperature, sealed, and passed through a pasteurizing tunnel.

Commercial preserves in the United States almost universally use frozen fruit as a raw material. Concentrated juices at 68° to 70° Brix are held in cold storage, conditions under which microbiological activity is absent for periods of a year or more. Concentrated fruit juices usually have had their volatile flavor essences removed during preparation. These flavors are returned to the batch just before the filling process.

Various types of instruments are used to evaluate the gel strength or viscosity of finished products. Consistency of preserves can be evaluated by spreading a given amount of product on a plastic plate engraved with concentric rings one centimeter apart. Readings taken at eight points around the circle are averaged to give a meaningful measure of product viscosity.

The Sunkist Growers Ridgelimeter has been the industry standard for the routine measurement of pectin grade quality. A jelly prepared under precise conditions is poured into a glass equipped with tape "sideboards," which allow the glass to be filled past its top. After the jelly has set a prescribed time, the tape is removed and the jelly extending higher than the top of the glass is removed with a cheese-cutting wire. The jelly is then carefully inverted onto a glass plate and the amount of height depression, or "sag," is measured and compared to the height of the glass container (not including the sideboards) in which the jelly was allowed to form.

Devices such as testers having controlled-speed plungers of defined configuration engage the jelly surface while being supported on an accurately tared scale. The plunger is allowed to descend onto the jelly until the pressure applied ruptures the surface. At this point, the scale reading reports the total pressure in grams and is recorded as the internal strength of the jelly.

Typical Formulas

Certain laboratory equipment and procedures will make jelly-making easier, more convenient, and more satisfying.

An electronic quick-tare laboratory scale of about 5,000-gram capacity should be selected. Such a scale should easily weigh the pan plus a stirring device, a thermometer, and all added ingredients without exceeding its capacity.

It is convenient to choose a finished batch weight of 1,000 grams. If a batch of 65 percent solids is made, the weight of sugar solids in the formula must be 650 grams. If the ingredients are carefully weighed into the pan, and the percentage of soluble solids contained therein are calculated, then the batch can be projected to end up at 65 percent solids when the total batch weight is adjusted to 1,000 grams. When the batch is hot, evaporation is rapid and manipulation to establish an accurate solids end-point is sometimes frustrating. With the selection of the 1,000-gram finished batch weight and knowledge of the grams of solids in the ingredients, then the net weight of the pan contents at the finishing point can be adjusted to 1,000 grams by either addition of water or further evaporation. The same principle holds true for any other batch size. It is more convenient to finish to a final weight than to cool a small sample to measure solids on a refractometer.

If the solids and the weight of all ingredients are known, the final solids can be calculated and the amount of evaporation or amount of water to add is known. To develop a formula starting with 100 grams of apple juice concentrate at 70° Brix, the equivalent weight of single-strength apple juice contained in the concentrate must be calculated. The government standard for single-strength apple juice is 13.3 percent soluble solids. Thus, dividing 70° Brix by 13.3° Brix shows that each 100 grams of original concentrate is equivalent to 525 grams ([70 ÷ 13.3] × 100 = 525) of single-strength juice.

For the 45:55 ratio of fruit to sugar, the amount of sugar to add to the formula is determined by multiplying 525 by 55, and dividing the result by 45. This calculation will give 641 grams of sugar in the present example. Now the individual formula weights and corresponding soluble solids weights can be set up for the batch.

Table 14-1. Prototype Formulation of Apple Jelly

Ingredient	Soluble Solids
100 g 70° Brix apple concentrate	70 g
641 g Sugar	641
4.0 g 150-grade slow-set pectin	4
6 ml 50 percent citric acid (≅ 6 grams)	3
751 g	718 g

The 718 grams of soluble solids (Table 14-1) will require a total batch weight of 1,104 grams to achieve a final concentration of 65° Brix. Since the batch weight is 751 grams before water addition, 353 grams of water are added to the formula. Mathematically, the level of soluble solids is divided by the final Brix, as a decimal, to determine total batch weight. Subtracting batch weight before water addition from total batch weight will equal the weight of water to add. This result shows that there will be plenty of water to disperse the pectin.

Another calculation that can be done shows the present mixture (before water addition) will have a soluble solids content of 95.6° Brix ([718 ÷ 751] × 100).

This calculation protocol reveals whether evaporation or water addition is needed, as well as the finished weight of the batch at 65° Brix.

A refractometer calibrated to the Brix scale will be accurate for percent solids by weight only for pure sucrose. Other sugars and ingredients such as citric acid will cause some small deviation since their light-bending properties are somewhat different. However,the legal definition of the soluble solids end-point is 65° on the Brix scale regardless of the nature of the other ingredients present.

For purposes of uniformity and to duplicate the traditional products, the formulas outlined in this section utilize sucrose. Technologists can use these formulas as a starting point and as a basis for learning the techniques of preserve and jelly making.

SUGAR/WATER JELLY

A good starting point to learn about making jellies is the following small batch of sugar/water jelly:

650 g sucrose

360 g water – 10 g to be evaporated

4.33 g 150-grade, slow-set pectin

2 ml citric acid solution (50 percent weight/volume) sufficient to reach pH 3.2

One gram of "150-grade" pectin will satisfactorily support 150 grams of sugar at 65 percent solids and a pH of 3.2. The quantity of pectin can be estimated on this basis. If the fruit portion contains a significant amount of pectin, less need be added.

1. Blend about 10 grams of the sugar with the pectin.
2. Disperse this mixture into the water, being sure that no lumps remain. A blender is a convenient tool for this.
3. Heat the pectin-water mixture until the pectin is completely dissolved (180° to 190°F). Do not add the remainder of the sugar until the batch is hot. Pectin will not hydrate completely when the solids are too high.
4. After adding the sugar, boil until the solids reach 65 percent as measured by a refractometer. (Of course, in this particular case the solids will be 65 percent when the batch weight

reaches 1,000 grams.) Be sure to precool the sample to avoid rapid evaporation on the refractometer prism. It is convenient to spoon some sample into a glass test tube and cool in an ice bath. Take note of the instrument temperature and refer to the Brix-temperature chart for the appropriate correction for temperatures other than 20°C. (See FDA Standard.)

5. Turn off the heat.
6. Check the pH and adjust to 3.2 with the acid solution.
7. Pour into jars and seal at once.
8. Allow to air-cool for 5 minutes and then water-cool to room temperature. Air-cooling allows the hot product to kill any mold spores that may be on the lid.
9. Yield equals 1,000 grams less amount of product lost as samples.

Now try a grape jelly made from concentrate at 68° Brix:

45 g unsweetened concentrate at 68° Brix

550 g sugar

3.6 g 150-grade slow-set pectin

5 ml citric acid solution (50 percent weight/volume) sufficient to reach pH 3.2

1. Mix the pectin with 120 grams of the sugar in a blender.
2. Blend the grape juice with the pectin-sugar mixture and heat.
3. At about 160°F add the remaining sugar and boil.
4. Cook until the solids reach 65 percent.
5. Check the pH and adjust to 3.2 with the acid solution.
6. Pour into jars, seal at once, and proceed as with steps 8 and 9 of the sugar/water jelly process.

A preserve or jam requires a somewhat different procedure. For a 47:55 ratio product:

470 g unsweetened fruit or purée, fresh or frozen. (If the fruit has been frozen with sugar, this sugar must be deducted from the calculated sugar amount.)
550 g sugar
3.2 g 150-grade slow-set pectin

2 ml citric acid solution (50 percent weight/volume) sufficient to reach pH 3.2

1. Mix the pectin with 107 grams of the sugar in a blender.
2. Combine the pectin-sugar mixture, fruit, and about half the remaining sugar; heat to a boil.
3. At about 180°F add the remaining sugar.
4. Cook until the solids reach 65 percent.
5. Check the pH and adjust to 3.2 with the acid solution.
6. Pour at 185°F or higher, seal, and cool as before. Yield: 900 to 925 grams depending on fruit solids.

Whole strawberries and cherries have a tendency to float. This problem can be minimized by cooking the early part of the batch to 68° to 70° Brix. This raises the Brix of the inside of the fruit to a value higher than 65°. Add water to return the batch Brix to 65° and fill as in prior experiments.

These examples will probably not give exact results the first time but will make the techniques more familiar. Subsequent batches made with appropriate adjustments will bring the final product on target.

The list of references will be valuable to the student who wishes to pursue the study of preserve and jelly making in greater depth and detail.

References

1. Code of Federal Regulations, Title 21 Part 150 "Fruit Butters, Jellies, Preserves, and Related Products."
2. Genu Pectins, PFW Division, Hercules Inc., Middletown, New York.
3. Grindsted Products Inc., Industrial Airport, KS.
4. Kertesz, Z.I. *The Pectic Substances.* Interscience Publishers, New York. 1951.
5. Nelson, Smit, and Wiles. "Commercially Important Pectic Substances." in *Food Colloids.* Edited by Graham, H. D. AVI/Van Nostrand Reinhold, New York. 1977.
6. Institute of Food Technologists. "Pectin Standardization: Final Report of the IFT Committee." *Food Technol.* 13, 496–500. 1959.
7. *Glass Packer* Magazine. 17 (7,8). 1938.
8. Rauch, G. H. *Jam Manufacture.* Leonard Hill Books, London. 1965.
9. Lopez, A. "Jams, Jellies and Related Products." In *A Complete Course in Canning.* The Canning Trade, Baltimore, MD. 1981.

10. Tressler, D.K., and Woodruff, J.G. "Jams, Jellies, Marmalades and Preserves, Candied and Glacéd Fruits, Fruit Syrups and Sauces." *Food Products Formulary*, vol 3., sec. 6. AVI/Van Nostrand Reinhold, New York. 1976.
11. Tressler and Joslyn. "Fruit Juices in Jelly Making." In *Fruit and Vegetable Juice Production*. AVI/Van Nostrand Reinhold, New York. 1954.
12. Meschter, E. E. "The Preserver's Stake in Uniform Pectin Grading." *Food Technol.* 3, 28–33. 1949.
13. *Pectin and Its Applications*, Unipectina, Bergamo, Italy.
14. *Chemistry and Function of Pectins*. Division of Agricultural and Food Chemistry Symposium, American Chemical Society, Washington, D.C. 1986.
15. *The Almanac of Canning, Freezing, Preserving Industries*. Edward E. Judge & Sons. Westminster, MD.
16. Drummond and Wilbraham. *The Englishmen's Food*, p. 58. London. 1958.

15

Sugar in Microwave Cooking

Triveni P. Shukla and Devendra K. Misra*

In no more than a decade, microwave cooking and heating applications have become a norm in household and commercial food processing in the U.S. (26,34). Similar developments in the area of consumer foods for the microwave are taking place in other developed countries (9,24,38). While microwave cooking of convenience consumer foods is an established application, new food processing applications of microwave energy will soon develop in the areas of industrial cooking, pasteurization, and sterilization (1,6,17,23,25,32). Many food process applications (27,31,32,42) have become successful manufacturing technologies, and the manufacturing economics are very favorable (11,20,27,32,36,43, 44).

Although not recognized as such, sugar with its unique dielectric properties (13,30) is a very functional food ingredient from the point of view of microwavable food formulations. The microwavable foods market in the U.S. alone has reached a $3 billion per year mark; it is projected to reach $10 to $15 billion by the year 2000. At least 20 percent of consumer foods, including bakery, candy, granola, pastry, confectionery, cookies, sweet snack, and prepared-mix products, are sugar-based formulations. The current market value (Table 15-1) of existing products and the ongoing developments in the area of new sugar-based convenience foods (Table 15-2) render sugar as a key "food dielectric." The role of sugar, as a special dielectric, will become even more important by the year 2000 when 1) 95 percent of the American households begin cooking in the microwave oven,

*Triveni P. Shukla is with F.R.I. Enterprises, New Berlin, WI. Devendra K. Misra is with the Department of Electrical Engineering, University of Wisconsin, Milwaukee.

2) 60 percent of the fast and prepared food categories become truly microwavable, and 3) the role of sugar-like dielectrics in food formulation is fully understood. The importance of sugar as a microwave-absorbing food ingredient will supersede its conventional role as a sweetener, plasticizer, and texturizing agent; its use as a control agent for microwavability and water activity will become exceedingly important in shelf-stable dry mixes, baked goods, bakery ingredients, preserved fruits, and sweet snacks. It may become a key process aid in commercial unit operations of pasteurization and sterilization. Although the dry-goods categories make up only 20 percent of the total U.S. food industry shipments ($342 billion per year in 1988), sugar is a key ingredient in these product categories. Therefore, the worldwide microwavable foods trend will have a positive impact on the growth of the sugar industry in the U.S. and elsewhere.

What follows is an account of microwave energy, its production in the microwave oven, its interaction with foods, its synergistic interaction in sugar-water binary systems common to foods, and the special role of sugars and carbohydrates in microwave cooking of consumer foods. Sufficient theoretical relationships are included in Appendix 6 as a reference in order to render applied considerations more meaningful and complete. A few recipes that exemplify the role of sugar in microwave cooking are also included. The details of dielectric theory are available in a number of publications (5,7,10, 13,15,46).

Table 15-1. Annual Shipment Value of Sugar-Based Food Products

	Billion Dollars					
	83	84	85	86	87	88
Bakery products	18.21	19.39	21.83	22.66	21.89	23.24
Confect.	7.17	7.78	7.91	8.08	8.43	9.12
Chocolate	2.26	2.48	2.59	2.66	2.77	2.56
Cookies	4.91	5.61	6.44	6.96	7.24	7.00

Source: U.S. Industrial Outlook, 1988

Table 15-2. Microwavable Food Product Introductions in the U.S.

Year	Number	Percent of Total
1986	278	4.5
1987	761	9.6
1988	962	11.7
Distribution of Products Introduced in 1988 (%)		
Bakery Products	33	
Baking Ingredients	22	
Candy Snacks	25	
Condiments	20	

Compilation, F.R.I. Enterprises (1989)

Microwave Energy

Microwaves, like light, are electromagnetic radiation: coupled electric and magnetic fields oriented at 90° to each other, they propagate at the speed of light—2.9999×10^{10} cm per second in a vacuum (Figure 15-1). Frequencies assigned for industrial, scientific, and medical use are given in Table 15-3. The microwave region falls between the frequencies associated with radio waves and infrared radiation (Figure 15-2). As is characteristic of all radiation, microwaves interact with matter by coupling to molecular structure.

Table 15-3. Industrial, Scientific, and Medical Use Frequencies (Federal Communications Commission)

Frequency	Frequency Range, ±	Wavelength
13.56 KHz	6.68 KHz	22.12 Km
27.12 KHz	16.00 KHz	11.06 Km
915.00 MHz	25.00 MHz	32.76 Cm
2,450.00 MHz	50.00 MHz	12.24 Cm
5,800.00 MHz	75.00 MHz	5.17 Cm
24,225.00 MHz	125.00 MHz	1.24 Cm

For all practical purposes, microwaves are very fast alternating current (2.45 billion cycles per second in a kitchen microwave oven versus 60 cycles per second for household alternating current). The energy in microwave current is only 4.5×10^{-34} watts–hour per quantum but is sufficient to polarize water and to move ions present in foods back and forth along changes in polarity. This property of microwaves is utilized for heating and cooking foods. The choice of 915 MHz for industrial and 2450 MHz for the kitchen microwave is based on optimum interaction of water, salt (ion), sugar, and other food ingredients with these frequencies.

Microwave

A microwave oven converts 60 Hz household electric power into 2450 MHz microwave power which is then irradiated into food.

It consists of a power supply, a magnetron that produces microwave radiation, a power coupler and waveguide, mode stirrers, a six-sided rectangular cavity of highly reflective surfaces including the door, and maybe a rotating turntable (Figure 15-3A). In func-

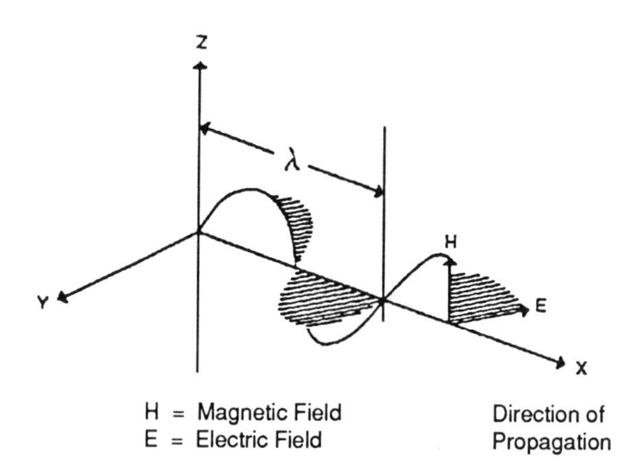

Figure 15-1. Plane wave electromagnetic field propagation. (Source: Ditchburn, R. W. *Light* Vol. 1. John Wiley 1963.)

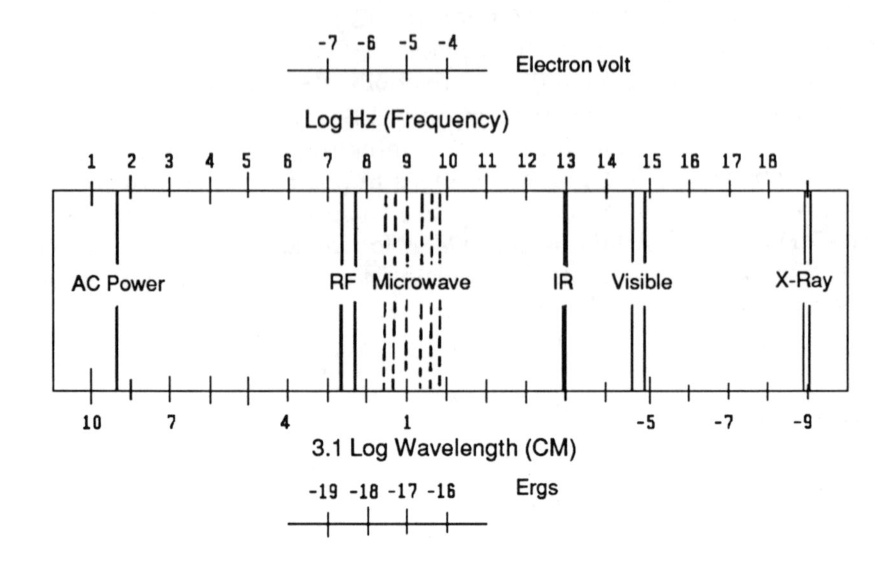

Figure 15-2. Electromagnetic Spectrum: Energy of microwave radiation. (Source: Shukla, unpublished, 1988)

tional terms, the oven should be an efficient generator of microwave power with mode mixer (for uniformity of electric field), sensor, and selector; it must have a technically sound, safe, and user-friendly timer and power control device.

The mode refers to the number of peaks along the travelling microwave (Figure 15-1) within the cavity; it depends on the cavity dimension. The knowledge of the mode distribution in the empty cavity is critical. The cavity is invariably rectangular for the ease of predicting mode density distribution. The length of the cavity is kept equal to one-half the free space wavelength or its multiple. The energy density distribution throughout the cavity is kept identical and, by design, microwaves ought to flood the cavity from all directions.

A 120 V, 60 Hz alternating current power supply is converted to a constant direct current after having been stepped up to 4 kV by a high-voltage transformer (Figure 15-3B). The transformer delivers a constant current over the normal range of line voltage. The filament shown in Figure 15-3B rises to electron-emitting temperature in no more than 1 to 2 seconds; the electrons, in effect, are the source of microwave power and the magnetron is an electron accelerator. The

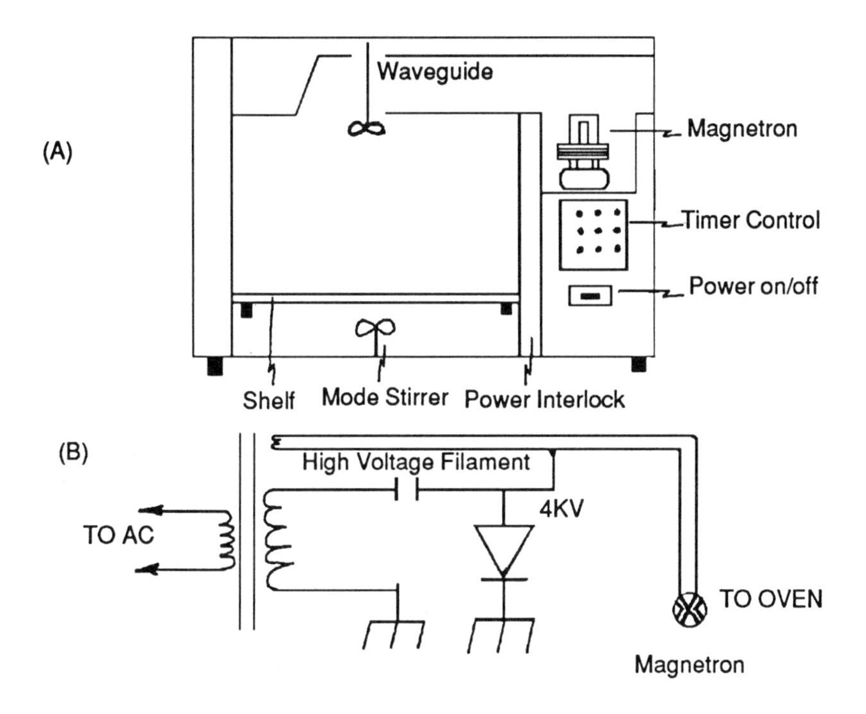

Figure 15-3. A typical microwave oven (A) and its power supply (B). (Source: *J. Microwave Power* E2:199. 1987)

oscillations typical of an alternating current are brought about by the magnetic field permeating an electrode assembly. The microwave power so generated is received by an antenna and finally delivered in the microwave cavity via the waveguide. Due to almost complete reflection at the walls, the power density pattern in the oven is characteristic of standing waves. By design, the output system of the magnetron is matched with food loads for efficient coupling or power transfer. The events of microwave power generation, loading effects, coupling efficiency, and energy transfer to food in a microwave oven are shown in Figure 15-4.

A 15 amp, 120 volt outlet allows a maximum of 750 watts of microwave power. High microwave power for institutional uses is also available: 1 kW at 20 amp, 120 V and 2 kW at 20 amp, 280 V. In practice, the power output and frequency depend on a number of factors: waveguide, food system, and the food in the cavity. The food can decrease or increase both power and wavelength. The microwave power generated equals the sum of power reflected back to the magnetron, power absorbed by the feed system, and the power absorbed by the food. Therefore, the size and the absorption characteristics of food are extremely important in regard to the actual power delivered during heating and cooking.

Although the number of modes increases with the volume, small dimensional changes in width, depth, and height can cause large mode changes. The magnetron sees mode changes by the mode stirrer as an effective change in cavity dimension.

Commercial microwave ovens vary in power from 400 to 750 watts and in cavity volume from 0.4 to 1.8 cubic feet. The feed system may have mode stirrers, rotating wave guides and antenna, a rotating turntable, and a bottom glass/ceramic plate. The latter acts as an absorber and protects the magnetron under no-load conditions. A combination of mode stirrer and turntable is good for proper mode distribution and uniform heating. An ideal microwave oven should receive microwaves from both top and bottom. Some ovens may include resistance heaters for browning and crisping foods in metallized containers at temperatures (350° to 400° F) common to conventional electric or gas ovens.

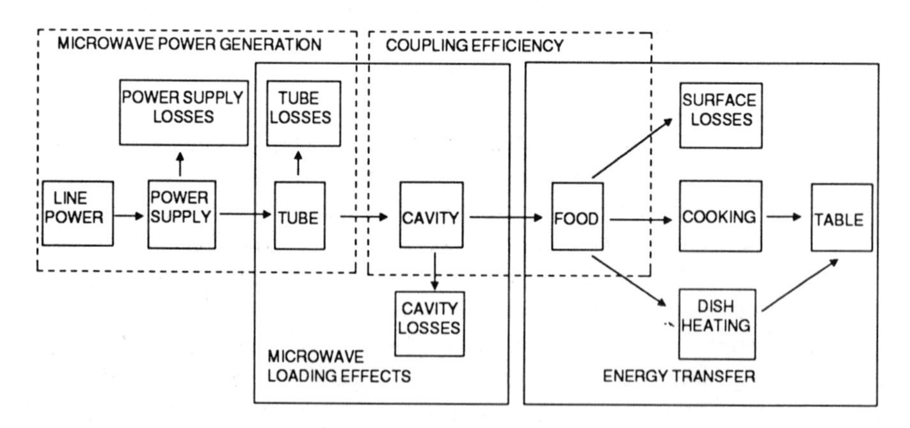

Figure 15-4. Power generation and transmission in a household microwave. (Source: *J. Microwave Power* E2:199. 1987)

However, kitchen microwave ovens have not been standardized with respect to critical engineering parameters that affect heating and cooking of foods. Most ovens operate at an electrical efficiency of 50 percent. A 15 amp, 110 V circuit in the kitchen has 1650 watts of power. At 50 percent efficiency, approximately 750 watts of microwave power is the upper output limit. This can be varied by varying the duty cycle (percent of the time the magnetron is on). The typical duty cycle for a household oven varies between 10 and 15 seconds and the magnetron takes less than 2 to 5 seconds to reach its electron emitting temperature.

Mechanism of Heating

Foods heat under the influence of microwaves due to energy dissipation caused mainly by the "orientation-polarization" of water (absorptive loss) and space charge polarization (I^2R loss) of food electrolytes; heat production is essentially due to interaction of energy with matter causing increasing motion and mobility.

Microwave power absorption occurs because of displacement and alignment (polarization) of water dipole and charges under the influence of a microwave electric field. The main displacements are due to dipolar (asymmetric positive and negative charge) water molecules and ionized salts whether organic or inorganic. Overall heating depends on temperature rise and the viscosity of the food. As water molecules in foods align and relax when coupled to a microwave frequency, heat results from motion and attendant friction. Heating due to ions further depends on their volume (hydrated or nonhydrated) as it influences motion in a medium of given viscosity. For a fixed frequency heating at 2450 MHz, the size of polar molecules other than water and the viscosity of food are important with respect to relaxation of interacting food constituents and, therefore, to heating (7,13). Relaxation in this context means reorientation back to the state without the electric field effects of microwave radiation.

The Food Materials

Microwave cooking is an excellent example of interaction of energy with moving charges and charged systems in foods and consequent production of heat in a given volume of food. Microwaves interact with food as an *electromagnetic* system. As a mixture of organic materials, it contains water, electrolytes, and other polar or apolar microwave-reactive substances. It often contains varying amounts of air trapped in it. As nonmagnetic composite matter, foods are mixed dielectrics. The electric flux density that develops in foods, when placed under the influence of microwaves, is proportional to field intensity. Since foods are noncrystalline, the microwave heating effect is homogeneous or isotropic.

As nonmagnetic matters, foods interact only with the electric component of the microwave, not the magnetic. Under the influence of microwave fields foods can store, absorb, and dissipate energy. The description and quantitative evaluation of these events requires a detailed understanding of the dielectric properties of foods.

The heating rate depends on amounts of water and volume of hydrated or unhydrated polar constituents of food ingredients—flavors, seasonings, emulsifiers, and polyhydroxy alcohols. Orientation polarization described earlier is largely due to free water (40). The polyhydroxy alcohols (sugars, carbohydrates, syrups) accentuate the absorption of microwave energy synergistically due to the orientation-polarization of water. This is true of sugar-water and alcohol-water mixtures in general (Figure 15-5).

The conduction effects (resistive heating) are due to food electrolytes—salts, organic acids, free amino acids and hydrolyzed proteins, leavening agents, mineral supplements, amphoteric lipids, small molecular weight glycolipids and glycoproteins, and short-chain fatty acids and amino sugars. These substances can influence the heating rate although present only in small (0.1 molar) quantities. Since it is the dissociated or ionized form of electrolytes that interacts with microwaves, the pH and ionic strength effects become critical and significant (10,22,39).

As mixed dielectrics (mixture of proteins, carbohydrates, lipids, electrolytes, and water), food compositions can support certain resonances and modes when placed within a microwave oven. This dielectric behavior of foods depends on their dimensions and dielectric properties and is modifiable by specific formulations.

Sugars in particular can be used to modify the food surface for browning and crisping. Each food product changes the wavelength of an incident microwave to a different extent. The food scientist has to rely heavily on the dielectric property of food ingredients in developing microwavable foods. Salt, polyhydroxy alcohols, and sugar are key ingredients in this regard (Table 15-4).

Table 15-4. Comparison of Complex Permittivities of Aqueous Solutions of Polyhydroxy Substances (2.4 GHz)

Substances	Real	Imaginary
Water	78	11
Methanol	22	13
Ethanol (95%)	11	10
Glycerol	7	4
Propylene glycol	7	6
Aqueous Solns. (50%)		
Methanol	50	16
Ethanol	43	18
Glycerol	46	25
Propylene glycol	34	22
Sucrose	42	18

Source: Shukla and Misra, unpublished data, 1989.

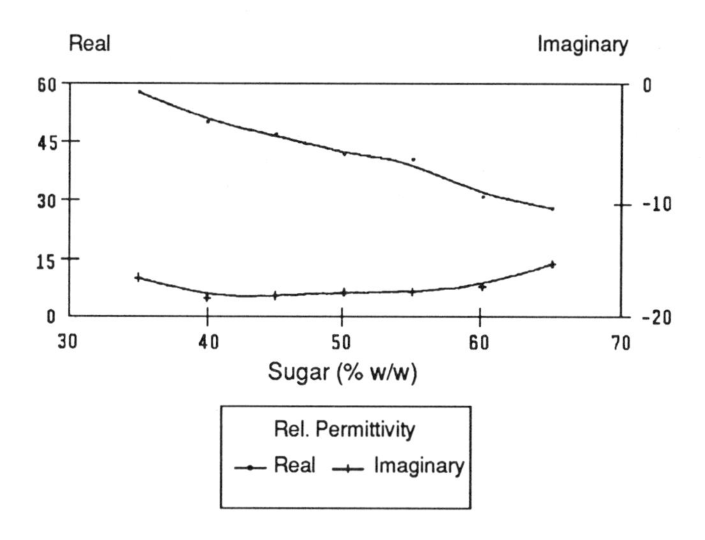

Figure 15-5. Variation in complex permittivity of aqueous sugar solution (65°F, 2.4GHz) as a function of concentration. (Source: Shukla and Misra, unpublished, 1989)

Foods are not easily predictable dielectrics because their properties depend on both frequency and temperature. The dielectric properties of foods change as they undergo heating to higher temperatures. The heating rate and heating uniformity is determined by shape, size, and density of the food. They can be heterogeneous and polyphasic, e.g., pizza, chunky soups, with respect to low and high dielectric constant or dielectric loss. The heterogeneities may originate from differences in structure and texture, porosity, thickness, moisture, density, polymeric network, viscosity, and water activity. Food ingredients may exhibit relaxations of different orders and electrical conductivities of different magnitudes involving various fixed- and free-charge systems. Therefore, formulation of a food product for microwave cooking must be based upon the dielectric properties of its major and minor ingredients along with point and/or surface charges.

In theory, however, we must recognize that, at a microwave frequency of 2450 MHz, the important phenomena in relation to heating are the orientation-polarization of water (the permanent dipole) and the space charge polarization of ions (dissociated salts). Significant differences due to major and minor ingredients alike via their relative contribution to a food's microwave energy absorption behavior can have demonstrable effects on the rate of heating in a given temperature domain. Pure water is ionic to an extent of 33 percent due to its dipolar nature. Sugars modify its behavior. What happens to the dielectric property of a 50:50 mixture of sugar-water or alcohol-water is an unknown (30). These mixtures are often twice as absorptive as the individual components of the binary mixture (Table 15-4). This interactive behavior offers a definite advantage in microwave cooking of sugar-based food formulations.

The Dielectric Properties of Foods

Dielectric constant and loss factors, the two components of complex permittivity, are two main properties of foods relevant to food formulated for microwave cooking. The two properties together determine the rate and efficiency of microwave heating and cooking. A number of other electrical and physical properties (specific heat, density, viscosity, and thermal conductivity) become important in calculating and interpreting heat absorbed, depth up to which microwave heating occurs, and the degree of microwave power conversion into heat. The following is a brief description of the relevant properties.

Table 15-5. Dielectric Activity of Major Food Constituents at Microwave Frequencies

Food Constituents	Relative Activity
Bound water	Low
Free water	High
Protein	Low
Lipid	
Triglycerides	Low
Phospholipids	Medium
Complex carbohydrates: starch	Low
Monosaccharides	High
Associated electrolytes	Low
Ions	High

Source: F.R.I. Enterprises: Tech. Bull. 86-002

Complex Permittivity. The complex permittivity of a dielectric (food) is described in terms of dielectric constant and loss factor (Equation 2, Appendix 6) in a frequency range where phase differences, polarization, and energy transfer occur simultaneously. Phasor representation of complex permittivity is shown in Figure 15-6.

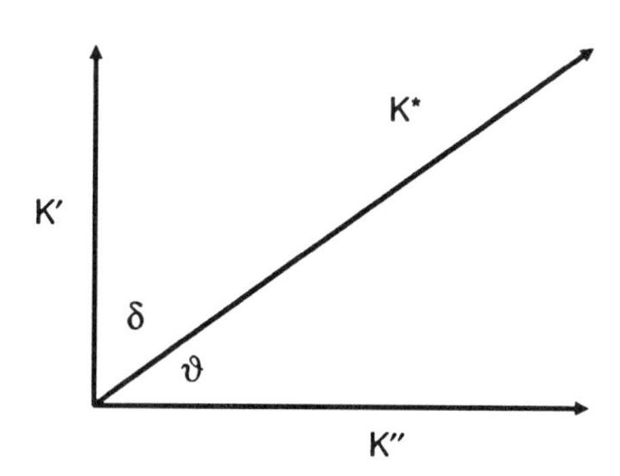

Figure 15-6. Phasor representation of complex permittivity. (Source: Shukla and Misra, unpublished, 1989)

Dielectric Constant. The dielectric constant is the real part in the mathematical expression of complex permittivity. In a physical sense, it is the proportionality constant that relates the force of attraction between two charges with the product of the two charges divided by the square of the distance between them. Also, it equals the ratio between the capacitance values of a capacitor with food and vacuum as the dielectric respectively. In terms of microwave heating, dielectric constant of a food material determines its ability to reflect microwaves. Foods with high dielectric constants reflect more and couple less of the microwaves falling on them, e.g., water reflects more than oil.

Loss or Dissipation Factor. This is the imaginary part of complex permittivity. "Loss" refers to loss of microwave power to food by absorption. The absorption of microwave radiation occurs because foods are not ideal capacitors. The quantitative measure of loss is the product of the dielectric constant and the tangent value of an angle that equals 90° minus the loss angle. In practice, the loss factor determines how well a food heats by microwave. High loss factor values are associated with high heating rates or high absorption of microwaves (See Table 15-5). Sugar and sugar-like ingredients improve heating of foods in microwave ovens in two ways: the lower dielectric constant aids coupling and the higher loss factor permits more absorption than water.

Loss Tangent. This is the ratio between the loss factor and dielectric constant (Equation 3, Appendix 6). For any electric field and radiation, it determines the penetration depth: the distance up to which microwave power is absorbed as it is transmitted through food (Appendix 4). See Equations 5 and 6 in Appendix 6.

Intrinsic Impedance. This parameter is defined as the ratio between the electric and magnetic fields (Equation 4, Appendix 6). It determines how well the microwave power couples with the food being heated in a microwave oven. For maximum coupling, the two fields must change in the same ratio as the microwave travels from air to food. This can happen only when the intrinsic impedance of foods is the same as that of air. This is not the case (Appendix 5). Olive oil has a value of 218, which is closer to that of air with a value of 377. Water has a value of 18. Since the mismatch between water and air is higher than between olive oil and air, the latter couples more microwave power.

For maximum power coupling (absorption of most microwave power), electric and magnetic fields must change in the same ratio. A poor coupling of microwave power by most foods can be visualized from their intrinsic impedance values in comparison with the 377 ohms of free space: high-moisture foods—50 ohm and dry foods—200 ohms. Dry foods exhibit higher coupling efficiency than moist foods.

Other Electrical Properties of Foods

Three properties described in this section affect the distribution of electrical energy in foods irradiated with microwave energy: attenuation factor, penetration depth, and power absorption.

The *attenuation factor* is defined in terms of a food's dielectric constant and its loss tangent (Equation 5, Appendix 6). It varies with time and position during the heating cycle as a function of frequency and local temperature gradient. The effect of temperature is indirect because the major dielectric properties are temperature dependent. A pictorial view of attenuation is represented in Figure 15-7. The attenuation per radian in a polar coordinate is called *index of absorption.*

The *penetration depth* is the reciprocal of the attenuation factor (Equation 6, Appendix 6), which is the depth below the food surface at which the electrical field strength is $\frac{1}{e}$ th that of the field strength in the free space. The distance at which it is 50 percent that of the free space (Equation 6, Appendix 6) is of more day-to-day practical value. The penetration depth can be approximated as

$$D_e = \frac{\lambda_o \sqrt{K'}}{2\pi K''}$$

A low penetration depth simply means high loss factor or absorption. It can be calculated for a microwave of 2450 MHz in space if the dielectric constant and the loss factor are known.

Microwave *power absorption* by food depends on the distance from the surface and the attenuation factor; the phenomenon follows Lambert's law (Equation 7, Appendix 6). The penetration depth (distance up to which microwave power is absorbed effectively) is inversely proportional to the frequency and loss factor but directly proportional to the dielectric constant. Low frequencies have high penetration power. For example, the 2450 MHz household microwave oven will heat foods to a shallower depth than industrial microwave ovens at 915 MHz frequency.

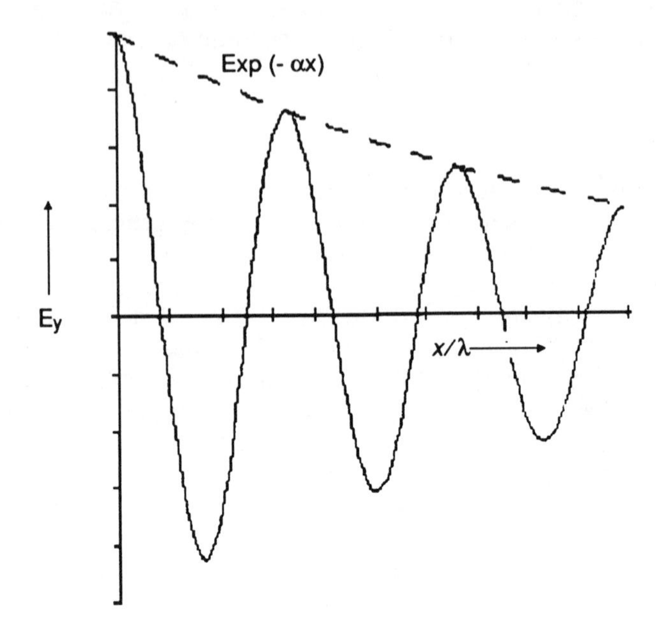

Figure 15-7. Schematic representation of the exponential decay of microwave radiation in absorptive media. (Source: Shukla and Misra, unpublished.)

Microwave Power Absorbed by Foods

The power absorbed per unit volume of food has the following relationship (Equation 8, Appendix 6):

$$P = kfE^2K''$$

The parameters f and E are designed into the microwave oven. Thus, loss factor is the key electrical property of foods that determines the rate of heating.

The common belief of "inside-out" heating of foods has meaning only when the largest dimension of the food is smaller than the wavelength in free space (12.25 cm). Although conditional for a number of reasons, this relationship does describe the physical process of microwave power conversion into heat.

A household microwave oven, being an enclosed cavity, traps the electrical field within it. It gives rise to a standing wave pattern that causes nonuniform electric fields with points of maximum and minimum energy concentrations. The turntable and mode stirrers are designed to create a more uniform power distribution. But the effects of different foods on effective wavelength is difficult to manage as both the dielectric constant and the loss factor are temperature dependent.

Based on the temperature rise measurements, the rate of heating can be calculated by Equation 9 in Appendix 6.

Interaction of Microwave Energy With Food Constituents

Food solids (proteins, carbohydrates, gums and hydrocolloids, and a majority of lipids) are dielectrically inert (Appendix 4). Their dielectric behavior is related to composition, water activity, structure, and their interactions with water and electrolytes (21). The dielectric activity of foods at high frequencies is due to water relaxation and ionic conductivity. In general, lipids and colloids derive their activity from volume exclusion. The dielectric activity of proteins may result from two sources: (a) high activity due to charge effects of 1) ionization of carboxyls, sulfhydryls, and amines, 2) hydrogen and ion bonding as affected by pH, and 3) net charges on dissolved proteins; and (b) relatively low activity due to relaxation and conductive effects. Both of these activities can be very important for hydrolyzed protein containing high levels of free amino acids that are often part of high-sugar foods (40,41). The microwave activity of proteins in high-moisture foods, however, is largely due to their colloidal nature. They modify the aqueous phase when their charged surfaces interact with ions (22). Similarly, starches and gums in high-moisture foods are expected to be dielectrically active mainly because of volume exclusion and hydrogen bonding effects (10). On the contrary, soluble carbohydrates, such as sugar (sucrose) and glucose, and other polyhydroxy food ingredients exhibit high dielectric activities in aqueous mixtures (30,31). Figure 15-8 depicts a schematic summary of such effects for foods viewed as a mixture of food solids in an aqueous environment.

Orientation-Polarization of Water. The dielectric constant of water (25° C, 2450 MHz) is 77. The observed values of dielectric constant and loss factor in the 2450 MHz region are temperature dependent. The frequency response is characterized by a dispersion phenomenon whereby the dielectric constant decreases logarithmically with the frequency from 80 in the static region to 5.5 in the optical region (visible light frequencies). But the loss factor increases from a negligible value in the static region to a maximum at the critical frequency and then decreases to a negligible value in the optical region.

In the dispersion region (the microwave region, Figure 15-2), dipole rotation is out of phase with the applied field, and it lags behind field reversals to an increasing extent as the frequency increases. The dielectric loss increases to a maximum at the critical frequency and is related to a characteristic relaxation time by the following relationship:

$$\tau = \frac{1}{2\pi f_c}$$

where $f_c = \frac{V_c}{\lambda_c}$,

where V_c = speed of light in a vacuum, and

λ_c = critical wavelength.

Relaxation time is defined as the time taken by the maximally aligned or ordered state of a dipole (e.g., water) to disorient to $\frac{1}{e}$ th level of order on sudden removal of the applied field.

The conduction behavior of water is the mirror image of its dispersion behavior. In contrast, however, the dispersion phenomenon is limited to one decade for the dielectric constant; it extends to two decades for the loss factor (Figure 15-9). Both the static dielectric constant and the critical frequency of water decrease with increasing temperature.

Water bound to food polymers (starch, gums, and proteins) relaxes well below the microwave frequencies. At 2450 MHz, bound water has almost ice-like dielectric behavior. The stronger the bonding forces between water and protein or carbohydrates, the smaller the values of the dielectric constant and loss factor. Since the bonding forces become weak at high temperatures, the dipoles may then be able to orient (Appendix 7).

The ion-like behavior of the water dipole presents a major means of converting electrical energy to heat in microwave cooking, and the management of "water activity" in foods is a primary means for formulating microwavable foods. The relative dielectric activity of major food constituents is listed in Table 15-5. Polarization of water and ion displacement are the two key phenomena responsible for food heating by microwave radiation.

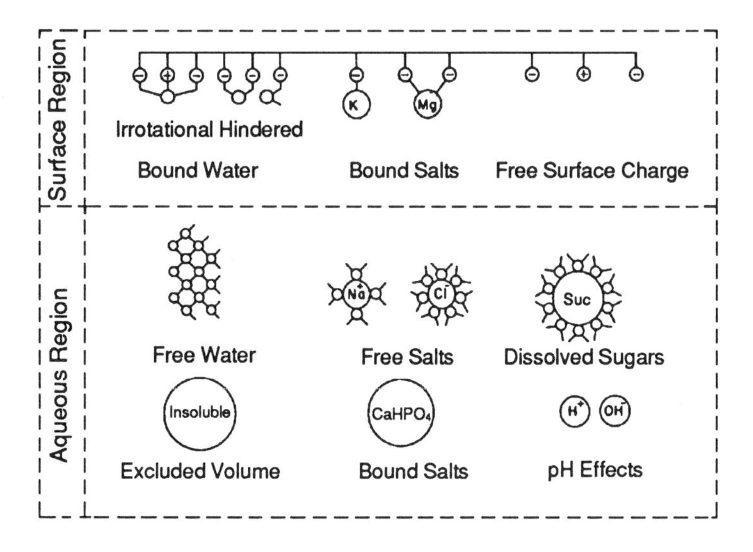

Figure 15-8. A model of interactions of microwave energy with foods. (21)

Behavior of Salts and Ions. Dissolved and dissociated salts have two major effects on the dielectric behavior foods: hydrated ions, depending on their size, hydration number, and charge, depress the dielectric constant of water below that of pure water on one hand and elevate the loss factor due to conductive effects on the other. The ions act as charge carriers. They increase heating rate and reduce penetration depth. The effect on the loss factor is governed by the equivalent conductivity of the ion; it is both temperature and concentration dependent and varies with the frequency. These relationships are shown in Equation 11, Appendix 6. In practice, the loss factor can be determined from the equivalent conductivity measurement alone.

Ionic and dipole losses are the dominant components of total loss due to the aqueous phase of foods. At a fixed frequency, dipole losses decrease but ionic losses increase with increasing temperature. Thus, total loss of food products decreases initially with increasing temperature when dipole loss is the dominant component but then increases with increasing temperature when ionic loss becomes the dominant component. The higher the ionic conductivity of a food, the lower the temperature and the quicker the ionic loss begins to dominate.

Behavior of Sugars and Alcohols. Aqueous solutions of sugars and alcohols are unusual in their dielectric behavior (Figures 15-5 and 15-10). The loss factor profile with respect to the percent of sugar or glycerol in water (Figure 15-10) exhibits a maximum at 50 percent. Also, the maximum loss factor seems to be related to the carbon number in the sugar or alcohol molecules. This intriguing phenomenon has interesting implications in browning and crisping formulations. This behavior, in all likelihood, depends on hydroxyl-water interactions that stabilize water structure by hydrogen bonds. The observed synergy of dielectric effects between water and sugars and its dependence on a critical sugar concentration need in-depth study. In all cases, the pure compounds are highly miscible in water. They shift critical frequency toward the microwave region away

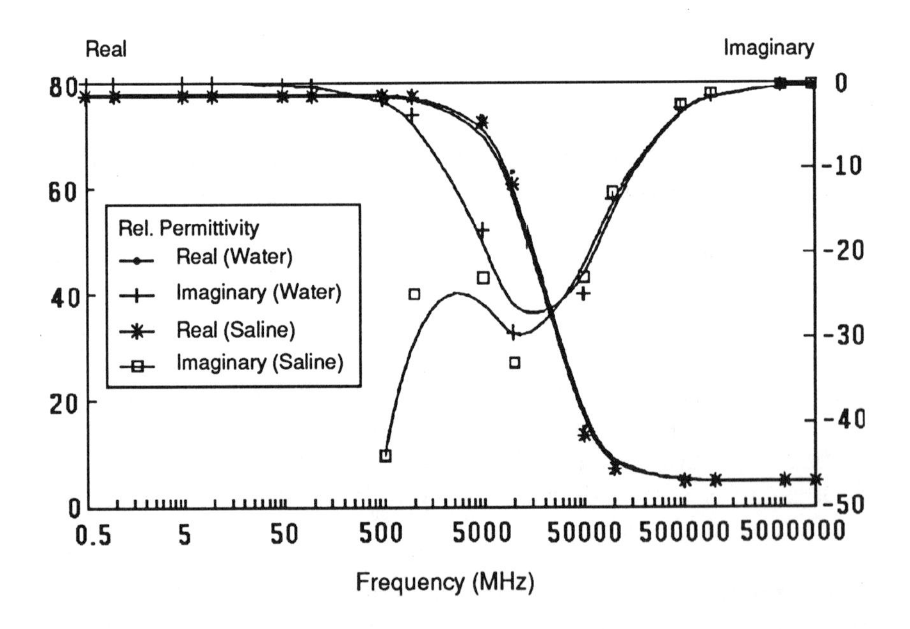

Figure 15-9. Dispersion behavior of water and aqueous salt solutions (ambient, 0.1 molar) with respect to real (dielectric constant) and imaginary (loss factor) parts of complex permittivity as a function of microwave frequency. (Misra and Shukla, unpublished, 1989)

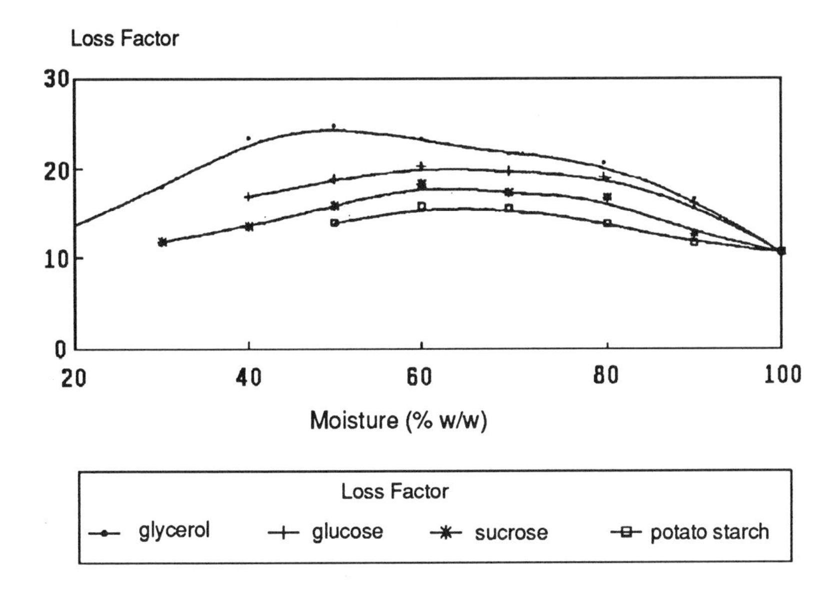

Figure 15-10. The loss factor dependence on concentration of sugar and sugar-like polyhydroxy food ingredients. (31)

from that of the pure components (See definition in Equation 10, Appendix 6).

This synergy in the dielectric property can be utilized in developing microwavable foods containing higher levels of sugar and polyhydroxy alcohols: syrups, confectionery items, sweet snacks, and candies. It is already used in formulating multilayered products for layer-by-layer selective heating. For instance, a topping can be designed to be heated in the microwave oven leaving the frozen dessert unheated.

The differences, although minor, in the loss characteristics of monomers, dimers, and polymers is also interesting from the point of view of interactions at the molecular level. Whereas a 5 percent sugar solution exhibits a negative temperature coefficient for the dielectric constant, the temperature coefficient for the loss factor is positive (Figure 15-11). A positive coefficient relates to fast heating to a high temperature. These findings were confirmed in recent work (39). Additional research work is in progress in order to understand the underlying phenomenon in detail. Recent dielectric constant and loss factor data are compiled in Table 15-4.

It is clear from the foregoing that the chemical composition of foods with respect to water, salts and ions, and sugar is the major determinant of their dielectric properties; proteins, starch, and gums can further modify them. Collectively, these ingredients in a formula can be varied to control the rate of heating, depth of heating, and efficiency of power transfer (coupling). The formulator needs to understand not only the individual ingredient effects but also the changes in the state and structure of water. Various schemes of water activity control in foods need critical analysis in terms of its direct and indirect effects on microwavability.

Factors Affecting Microwave Cooking

Foods cannot be predictably and uniformly heated in a microwave oven at 2450 ± 50 MHz by simply specifying their weight in relation to the power rating of microwave ovens. A number of variables that influence the heating of foods in a microwave oven should be evaluated and specified.

Moisture

The major permanent dipole in foods to which microwave energy is coupled (transferred) for heat production is water. Although the relaxation behavior of water in foods is complicated and poorly understood, high-moisture foods have a high dielectric constant; they heat fast due to a high loss factor. Their loss factor increases with moisture up to a limit and then decreases to 20 percent of the maximum value at water levels in excess of 80 percent (Figure 15-12). Usually the dielectric constant of a mixture attains a value intermediate to those of the pure components with the exception of sugar-water and alcohol-water mixtures as described in the previous section. Water activity, not moisture per se, affects both power coupling and power absorption. The concept of water activity and the technical criteria for effectively managing it are central to developing microwavable foods. The microwavability of high-sugar and intermediate-moisture foods is critically related to water activity. Even a low-moisture food may heat quickly due to its low specific heat.

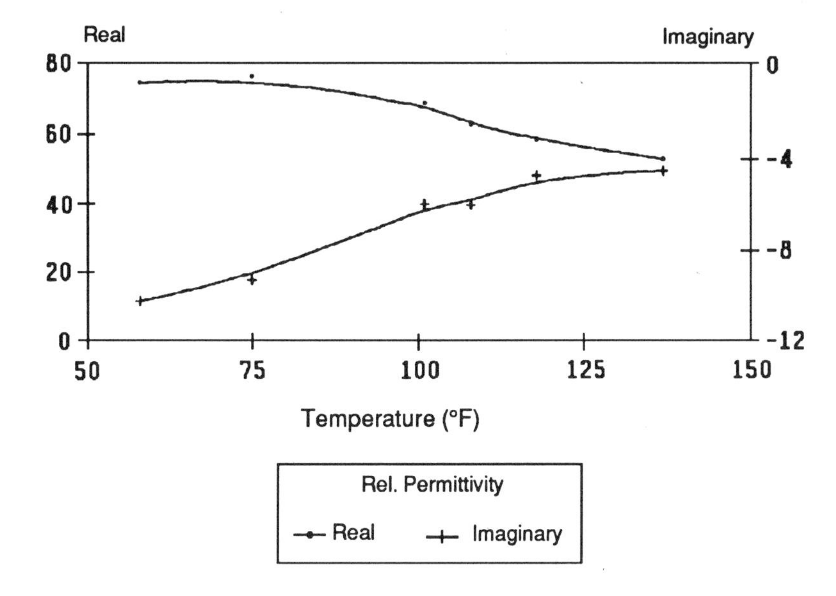

Figure 15-11. Temperature dependence (2.2GHz) of the real (dielectric constant) and imaginary (loss factor) components of complex permittivity of 5 percent sugar solution. (Source: Shukla and Misra, unpublished, 1989)

Density of Foods

Porous solids with entrapped air have a low dielectric constant because the dielectric constant of air is very low (1.00). It is known that both the dielectric constant and the loss factor increase with increasing bulk density (Figure 15-13) and that density is inversely related to the rate of heating (Equation 9, Appendix 6). Therefore, a low-density product exhibits a higher coupling efficiency and a higher rate of heating than a high-density food.

Temperature

Both dielectric constant and loss factor are temperature dependent. The loss factor variable is more important with respect to heating when its temperature coefficient is large and positive. This behavior can be effectively utilized for browning and crisping. A high and positive $\frac{dK''}{dT}$ (temperature coefficient of loss) value can be designed by formulation in order to achieve heating low-moisture foods to higher than 212° F. Sugar and salts are the key constituents for many browning formulations. Crisping can be brought about in similar manner.

Frequency

Although frequency is fixed at 2450 ± 50 MHz for the household microwave oven, the effective variations during transmission through foods during home cooking and industrial processing operations should be determined as a prerequisite for predicting uniform heating. This depends on the dielectric constant of the food as shown in Equation 12, Appendix 6. Water and high-moisture

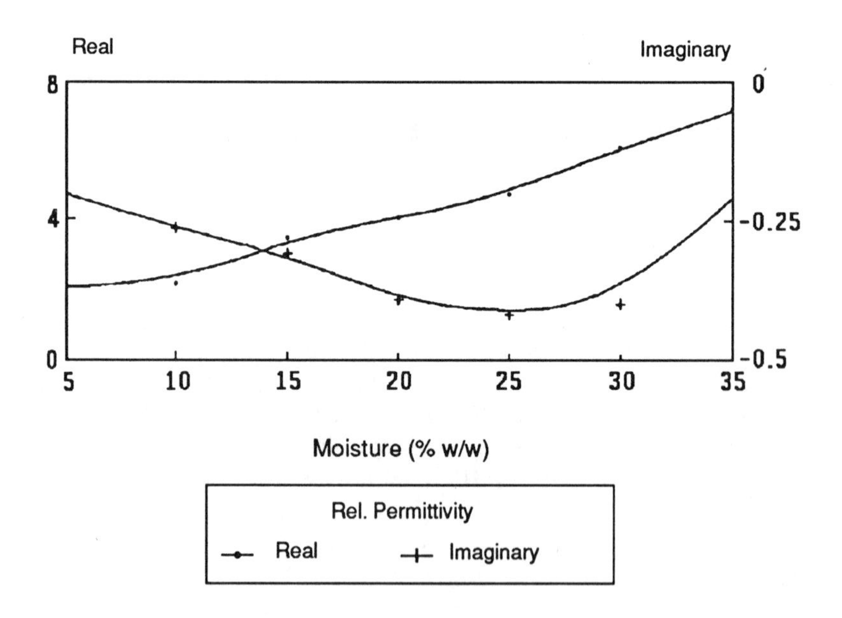

Figure 15-12. Dependence of the dielectric behavior of foods on their moisture contents. (Source: Shukla, F.R.I. Enterprises)

foods can reduce a 12.24 cm wavelength of 2450 MHz frequency to 1.37 cm ($\frac{12.24}{\sqrt{80}}$) as it travels through them.

Two other effects of frequency have to do with penetration depth and loss factor. At low frequency (long wavelength), the microwave penetrates deeper than at high frequency. Whereas the penetration depth in water is 1.4 cm for a 2450-MHz microwave, it is 3.7 cm at 915 MHz. The loss factor at 2450 MHz is 3 times greater than that at 915 MHz.

Geometry and Shape

Viewed from a microscopic point of view, particulate food solids have very complicated dielectric behavior because of the geometry of their individual particles and subparticles. But more important is the shape of the food product that is to be microwaved. The regular shapes (round, torus, and round-corner rectangular) are desirable.

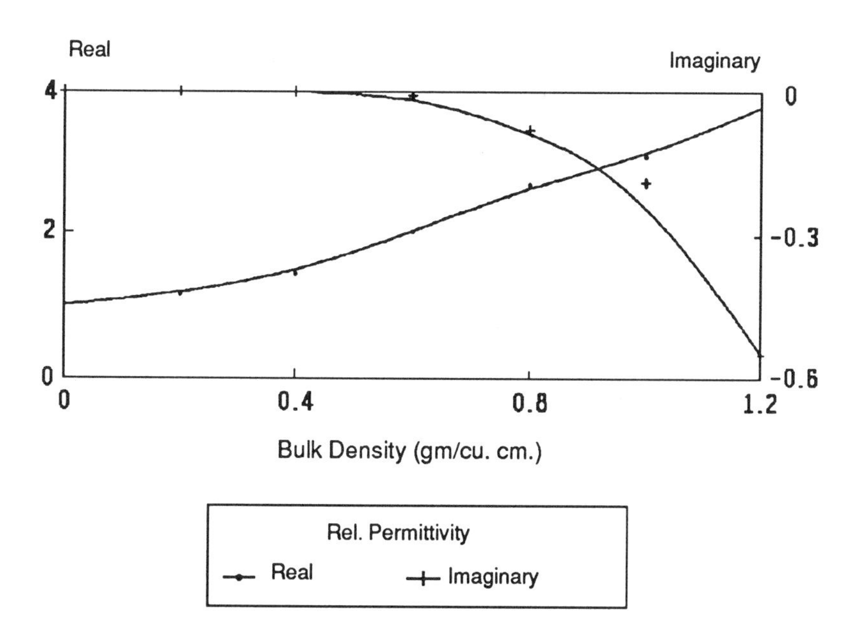

Figure 15-13. The relationship between components of complex permittivity of wheat flour and its bulk density. (Source: *J. Microwave Power* 19(1):56. 1984)

Any discontinuity in space, because of the corners and edges of the food, affects reflection and transmission of the microwave. At edges and corners, both internally reflected and incident power becomes available; it multiplies with the number of surfaces from which the reflection can occur or at which the microwave radiation can impinge (Figure 15-14). The overall effects can be visualized in terms of reflection from and transmission through foods, geometry-dependent energy concentration, and subsequent heating. Therefore, depending on composition and geometry, a region approximately 1 cm from the surface can have a very high rate of heating. The burning of corn chips is a good example.

Size, Dimensions, and Mass

Heating is often not uniform if a food product is larger in any dimension than the operating wavelength of the microwave oven. If it is the same size as the wavelength (12.24 cm), the center will heat to a higher temperature than the rest of the food. The food dimensions ought to be kept between 0.5 and 0.75 times the wavelength depending on the penetration depth, which is low for dry foods and foods with salt and sugar (Appendix 2).

Generally, the greater the mass of a food, the longer it takes to heat in the microwave. However, the efficiency of heating increases with an increase in mass. A single potato takes 4 minutes to cook, but two potatoes may take only 7 minutes (Appendix 1B).

Electrical Conductivity

As described earlier, the conduction component of the loss factor in most foods can have an overriding effect on heating rate. High loss factors associated with salts, hydrolyzed vegetable protein, sugars, and other polar and charged substances are responsible for two major effects—a decrease in penetration depth and, sometimes, high positive values of the temperature coefficient of loss. Careful consideration of both effects is necessary for a good microwavable food composition.

Although foods of high electrical conductivity permit only low coupling efficiency, this property can be used for surface heating, selective heating, shielding, and controlling the rate of heating in multilayered foods. Additional variables to be considered simultaneously are pH, ionic strength, ion size, ion hydration number, and ion valence. Addition of salt (sodium chloride) increases rate of heating and reduces penetration depth.

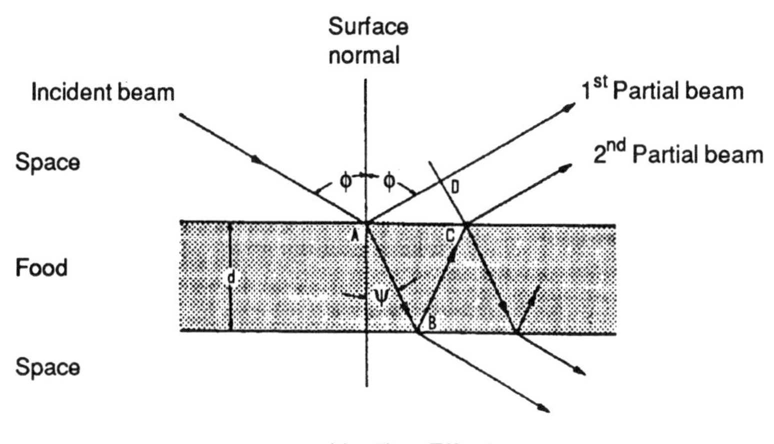

Heating Effects

Edge:	Energy in two dimensions.
Corner:	Energy in three dimensions.
Surface:	Fast heating at high salt and moisture.
Center:	Fast heating if the food's (load's) dimension is the same as the wavelength in free space.
Hot Spots:	High Voltage Standing Wave Ratio Low load.
Differential Heating:	Variation in the dielectric and physical properties of the food.

Figure 15-14. Transmission and reflection of microwave energy through foods—effect of geometry and dimension. (Source: Ditchburn, R. W. *Light*, Vol. 1, John Wiley 1963.)

Specific Heat

Heat capacity, like density, is inversely related to microwave energy absorbed by foods; it is, therefore, a key criterion to be considered in designing microwavable foods. The data in Appendix 3 show that its effects are significant as one compares low- and high-moisture foods. There are two food ingredients that affect its specific heat significantly: water and air. These effects relate to their specific heats. A third component is fat. Fat and lipid effects are significant when fats are used in high proportions. Therefore, heating rates can also be controlled by specifying moisture, fat, and extent of whipping. High-fat foods benefit from high coupling efficiency due to a closer impedance match with air in the microwave oven.

Thermal Conductivity

Once a food is microwave-heated, the subsequent temperature equilibration is dictated strictly by the thermal conductivity and heat transfer rates in the food. When cooking large food items for periods of 5 to 10 minutes, thermal conductivity may affect heat distribution during cooking. Determination of serving time after cooking can be made intelligently only with the knowledge of the thermal conductivity of foods. This is very true in the case of multicomponent entrées with more than one type of food, each heating at a different rate. The consideration of thermal conductivity effects in designing pasteurization and sterilization processes (high- and low-power treatment in increments) is critical (23,25,33). Proper design takes advantage of temperature equilibration arising from the thermal conductivity of foods; its effects are largely indirect.

Viscosity

Viscosity of foods affects the relaxation time (rate of unorientation) of dipoles under the influence of an applied microwave field. As described before, it is the time during which the maximally oriented state has exponentially decayed (unoriented) to the $\frac{1}{e}$ extent (5). The relaxation time is related to the viscosity of food by the Debye expression (Equation 10, Appendix 6).

$$\tau = V \frac{3\eta}{kT}$$

A high-viscosity food delays relaxation. Therefore, two foods of substantially different viscosity heat at two different rates. For illustration, viscosity of water (1.0 centipoise) is far lower than that of most food products; its relaxation time is approximately 10 picoseconds at room temperature. Viscosity of formulated foods can be as high as 3000 centipoise which, combined with the effective volume of proteins and carbohydrates, can increase the average relaxation time of food dipoles to the nanosecond range. For example, the relaxation time for bound water with D-glucose and starch is 20 compared to 1 nanosecond for free water.

Miscellaneous Effects

A number of other factors such as effective field differences in the food and oven, reduced power due to reflection in the microwave oven waveguide, level of impedance mismatch, and other radiation effects determine heating pattern and rate of heating. All effects have to do with microwave power. Therefore, all preceding variables should be optimized during food cooking in a microwave oven of a given power rating for an intended level and depth of microwave energy absorption.

The following is a list of problems in microwavable food product development owing to the interdependent effects of the factors described above:

a. Problems related to microwave oven: nonuniform microwave field, lack of uniformity between ovens, and surface cooling of foods due to evaporation;

b. Problems related to interaction of foods with the microwave: changes in loss characteristics with temperature; frozen versus thawed foods; lack of tolerance to rapid heating; deep-heating effects on bulk volumes in comparison to conventional oven heating where heating proceeds from the surface to the center; unusual temperature, pressure, and moisture gradients; and multicomponent (entrées) and multiphase (soup) foods;

c. Problems related to utensils and packaging materials.

Solutions to these problems can come only from a knowledge of the mechanism of microwave heating; an analysis of the thermal requirement of the food to be heated or cooked; designing a product that can tolerate high heating rates; using a low-power setting for cooking or reconstituting complex foods; assessing and optimizing loss factor, specific heat, and thermal conductivity of foods; choosing a proper shape, size, and material for the food container; and testing in ovens of different power specifications and designs.

Special Role of Sugar

Sugar has the special property of forming a permanent dipole when hydrogen bonded with water, which can be utilized in designing microwavable foods. In the aqueous phase of foods, it displays synergistically high loss characteristics with a positive temperature coefficient of loss up to a critical volume-fraction of water in foods (39). The two behaviors can be selectively utilized to increase the rate of heating, to do surface heating only, and, in combination with other absorptive materials of a high positive temperature coefficient, to formulate browning and crisping compositions. This area requires serious fundamental research in order to pinpoint the mechanisms underlying the dielectric behavior of mixtures of sugars and complex carbohydrates in water in terms of relaxation time, dipole character of the structural elements of the hydrogen bonded mixtures, and the molecular symmetry of the dissolved sugars.

Sugar is effectively used in both surface heating and creating a high loss shield that prevents the next layer of food from heating, e.g., topping in a plastic ice cream package. Also, sugar is used to moderate the dielectric constant of immersion water to the same value as that of food processed by microwave sterilization. This avoids uneven heating. The role of sugar as a process-aid ingredient is very significant in relation to microwave pasteurization and sterilization.

Major Applications

Major uses of microwave power in the food industry stem from the ease and speed of drying, cooking, and heat processing. This results in reduced process time and improved quality and yield. In-package microwave processing is possible in an open tubular cavity. Most commercial installations and operations offer the economies of space, labor, and overall manufacturing. Food processes most amenable to microwave processing are blanching; thawing of butter and meat; rendering of lard; finish drying of pasta, onions, crisps, and biscuits; pasteurization of bread; and reconstitution of prepared foods (cakes, french fries, entrées, etc.) in the food service sector.

Tempering or thawing of meat is the most widely used industrial practice with precise control over heat input and the resultant product temperature. Both 915- and 2450-MHz systems have been installed. In-carton thawing is very practical in a 915-MHz (high penetration depth) oven.

Although economically impractical for complete drying, microwave finish drying during the falling rate drying period is very advantageous with respect to process time, improved sanitation, and product quality. This is the portion of drying where both heat and mass transfer are limiting. It is employed for pasta, sugar, hops, and many textured food products. Fruit juices and other concentrates are dried to instant powders in vacuum microwave driers.

Microwave ovens at 2450 MHz are used in baking applications, although often in combination with conventional ovens. The process reduces baking time with no loss of quality.

Food products, when packaged in nonmetallic microwavable transparent containers (glass, plastics), can be pasteurized and sterilized successfully (17). The major problems in microwave sterilization are nonuniform heating and sublethal sterilization of particles larger than 2.5 to 3.0 cm in a 2450-MHz oven.

Bread and precut potato are good examples of in-package pasteurization. Food sterilization is also feasible. A number of processes have been developed. The Swedish immersion process and the Japanese keypack technology of sterilizing hot-filled acid-fruit products have received much attention. The control of surface temperature, hot-spot cooling, uniform heating, and overpressurization are the key advantages in addition to the economy in overall manufacturing.

Aseptic packaging, pasteurization of fish products, finish drying of snack foods, and process heating are in the development stages. Since sugar is an accepted ingredient of many consumer foods, it is the best candidate to be used for modifying the dielectric behavior of foods and food process aid compositions in microwave heating, cooking, and controlled thermal processing applications.

Selected Recipes

The recipes listed in this section are known to yield good quality microwaved foods in a household kitchen. The instructions are based on expected heating rates and attainable temperatures during cooking and baking. In some cases, containers with the provision of resistive heating are used and, in many cases, special effects of baking soda, fat, proteins, and sugar are sufficient to bring about desired temperature effects.

CHOCOLATE CAKE

Flour	2 cups
Unsweetened cocoa powder	2⁄3 cup
Baking soda	1 1⁄4 teaspoons
Baking powder	1⁄4 teaspoons
Water	1 1⁄3 cups
Sugar	1 2⁄3 cups
Eggs	4
Vanilla	1 tablespoon
Mayonnaise	1 cup

1. Use an 8-inch microwave cake dish and line it with a microwave towel.
2. Mix flour, cocoa, baking soda, and baking powder. Separately, beat sugar and vanilla for 3 minutes to a light and fluffy consistency and then beat in mayonnaise at a low speed. Add the flour mixture in four installments while agitating and mixing well.
3. Pour into pan. Cook at medium setting for 4 minutes, rotate, and cook another 3 to 5 minutes at high power setting until it begins to set on the sides.
4. Cool on a flat surface and then invert on a plate.

CHOCOLATE FROSTING

Butter	6 tablespoons
Cocoa powder	3⁄4 cup
Confectionery sugar	3 1⁄2 cups
Milk	6 to 7 tablespoons
Vanilla	1 teaspoons

Heat butter to soften. Add cocoa, sugar, and milk. Beat until spreadable. Blend in vanilla and spread on the cake.

CARROT CAKE

1. Combine 2 cups flour, 2 teaspoons ground cinnamon, 1 1⁄2 teaspoons baking soda, 1 teaspoon ground nutmeg, and 1⁄2 teaspoon salt and set aside.
2. Combine and blend 3 cups grated carrot, 1 ounce crushed pineapple, 1 cup shredded coconut, 1 1⁄2 cups sugar, 1 cup vegetable oil, and 1 cup chopped nuts.

3. Add 1 and 2 together and mix well.
4. Turn into a 12-inch ring (Bundt) pan. Cook on high power for 11 minutes. Turn if necessary. Cool and loosen sides. Invert.

FROSTING

Sugar	3/4 cup
Baking soda	1/2 teaspoon
Buttermilk	1/2 cup
Butter	1/4 pound
Corn syrup	1 tablespoon
Vanilla	2 teaspoons

Combine all ingredients except vanilla. Cook at high power for 2 minutes, stir well, cook another 4 to 5 minutes until slightly thickened, and then stir in vanilla. Add to the cake.

CHOCOLATE MOUSSE BROWNIES

1. Combine 2 ounces baking chocolate and 1/2 cup butter and cook at high power for 1 1/2 minutes to melt. Stir well.
2. Beat 2 eggs and add 3/4 cup sugar, 1/2 cup flour, 1 teaspoon baking powder, 1/2 teaspoon salt, 1 teaspoon vanilla, 1 cup chopped walnuts, and 1 cup chocolate. Blend well.
3. Add the chocolate mixture from Step 1.
4. Turn into 9-inch pie or quiche dish. Cook on high power for 6 minutes. Rotate after the first 3 minutes of cooking.

MOUSSE

Semisweet chocolate	4 ounces
Baking chocolate	1 ounce
Marshmallows, large	15
Milk	1/2 cup
Unsweetened cocoa powder	1 tablespoon
Vanilla	1 teaspoon

Combine all ingredients except vanilla. Cook on high power for 2 to 2 1/2 minutes. Stir once so that marshmallows are melted. Stir in vanilla. Let cook and thicken. Spoon it on the brownies. Refrigerate if not served immediately.

PECAN PIE

Unsalted butter	½ cup
Whole pecans	2 cups
Sugar	1 cup
Light corn syrup	1 cup
Eggs, lightly beaten	3
Vanilla	1 teaspoon
Pie shell, prebaked, 9-inch	1

Melt butter in 2-quart pan on high power. Add nuts, sugar, corn syrup, eggs, and vanilla. Blend well and pour into the pie shell. Cook the pie at medium high until the center is set—11 minutes. Turn after first 5 minutes if necessary. Serve chilled or at room temperature.

PEPPERMINT DIVINITY

Sugar	2 ⅔ cups
Light corn syrup	⅔ cup
Water	½ cup
Eggs, whites only	3
Peppermint flavoring	¼ teaspoon
Chopped almonds	1 cup
Red coloring	3 to 4 drops

Mix sugar, water, and corn syrup in 2-quart bowl and cook on high to 260° F to hard-ball stage (about 12 minutes). Beat egg whites until stiff. While beating egg whites, pour in cooked sugar syrup, color, flavoring, and almond.

Drop by tablespoon onto greased waxed paper. Cool and store in air-tight container.

PENUCHE

Light brown sugar	1 pound
Corn syrup	1 tablespoon
Butter	1 tablespoon
Evaporated milk	¾ cup
Vanilla	1 teaspoon
Chopped pecans or walnuts	⅔ cup

Use 2-quart casserole. Combine sugar, corn syrup, butter, and evaporated milk. Mix well. Cook on high power (700 watts) until the soft-ball stage (234 to 240° F). Stir in nuts and vanilla. Beat until thick and unglossy. Turn into 8- x 8-inch greased pan. Mark into squares and cut when cold.

CHOCOLATE FUDGE*

Sugar	2 cups
Unsweetened cocoa powder	½ cup
Half-and-half	⅔ cup
Unsalted butter	¼ cup
Chopped nuts	½ cup

Use a 4-quart casserole. Combine sugar, cocoa, and half-and-half, stir well. Add butter. Cook 6 to 8 minutes to boiling, stir well again, cook on medium power for 10 to 12 minutes to soft-ball stage (234 to 240° F), cool to lukewarm (110° F), beat by hand until thick, add nuts, and spread quickly on an 8-inch pan; if mixture hardens, add cream and continue beating. Cook and cut into squares.

COFFEE FUDGE

Add 2 tablespoons instant coffee with sugar to chocolate fudge recipe.

HARD CANDY

Sugar	3 cups
Light corn syrup	1 cup
Warm water	¾ cup
Flavoring oil	1 teaspoon
Food coloring	optional

Lightly butter or grease a cookie sheet. Use a 3-quart casserole. Add corn syrup, warm water, and sugar. Stir well. Cook 8 to 10 minutes on high until boiling. Stir well. Cook 12 to 15 minutes more to 290° F to hard-crack stage. Check after 7 minutes. Let stand until bubbling. Stir in flavoring and coloring. Pour on the greased sheet. Cool, break, and store in an air-tight container.

Variations: can be snipped, sugar coated, spiraled, or molded into lollipops.

ALMOND CRUNCH

Sugar	1 cup
Warm water	3 tablespoons
Melted butter	½ cup
Chopped roasted almonds	½ cup

---

*Recipes for Chocolate Fudge, Coffee Fudge, Almond Crunch, and Frosted Nuts adapted from *Mastering Microwave Cooking*, copyright ©1986 by Marcia Cone and Thelma Snyder. Reprinted by permission of Simon and Schuster, Inc.

Set aside an 8-inch greased square sheet. Use a 3-quart casserole. Combine sugar and water. Stir well. Add butter. Cook on high at 4 to 5 minutes until boiling. Stir well. Cook another five minutes until cracking point (290° F). Add nuts and mix when bubbling ceases. Spread on the greased sheet. Break when cool.

Variations: ½ cup semisweet chocolate chips can be added and spread evenly when molded. Also, can be topped with nuts.

FROSTED NUTS

Butter	2 tablespoons
Egg white	1
Sugar	½ cup
Unsalted nuts	1 cup

1. Use 2-quart microwave dish. Melt butter on high for 1 minute. Beat egg white until stiff and stir in sugar and nuts to coat evenly. Add all this to the molten butter.
2. Cook on high for 5 minutes until the nuts develop a white glaze. Stirring once when halfway through may be necessary.

These recipes call for reduced water content because its evaporation in a microwave oven is slower than in conventional procedures. Also, use of corn syrup is recommended for the control of crystallization and timing. Glucose and maltose syrups crystallize slower than sucrose. Specific temperature regimes are easily established.

Test	Temperature	Result
Soft-ball stage	235–240°F	Flattens
Firm-ball stage	245–250°F	Does not flatten
Hard-ball stage	250–255°F	
Soft-crack	270–290°F	Soft thread
Hard-crack	300–310°F	Brittle thread

High temperatures for various stages of sugar polymerization are possible with the help of sugar and fat, especially under low water conditions.

All recipes selected for this section are very absorptive (lossy), couple microwave power effectively, and make use of high temperature effects due to the presence of fat/oil, salt, sugar, and emulsified particulates. The recommendation of using high power stems from the lossy nature of a composition capable of heating quickly to temperatures as high as 310°F.

Appendix 1

(A) Indirect Assessment of Oven Power

Determine heating time to rolling boil for 8 oz (250 gm) water. Check approximate power of oven from the following table:

Wattage	Minutes to Boil
750	2.5
600	3.0
500	4.0
300	5.0
200	9.0
less than 100	does not boil

(B) Load Factor Curve

Most commercial foods range in mass from 100 to 250 gm. This is below maximum coupling efficiency. Although it varies from oven to oven, it should be determined as absorbed power.

$$P = \frac{(4.17)\ (\text{mass, gm})\ (T,\ ^{\circ}C)}{t,\ \text{minutes}}$$

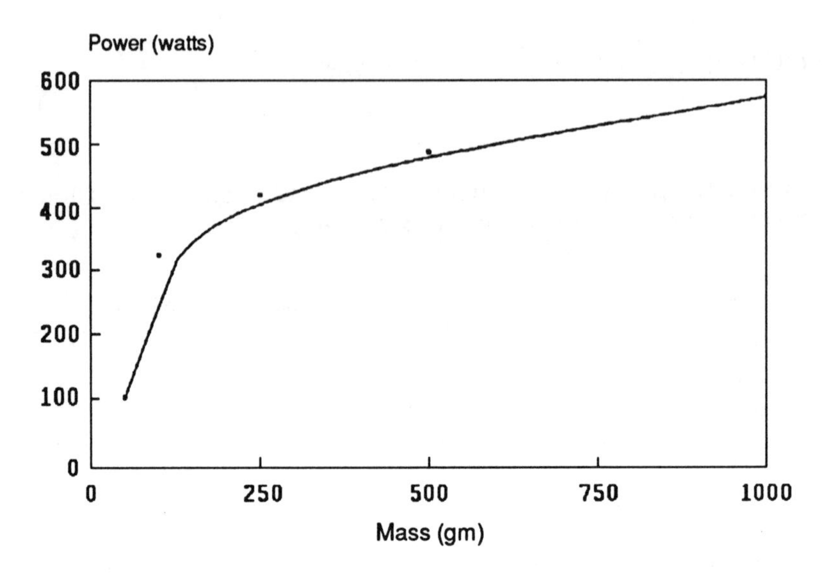

Figure 15-15. Load Factor Curve. (Source: F.R.I. Enterprises Tech. Bull. 86-002)

(C) Determination of Heating Time

Consider 80 percent efficiency in a 1000-watt oven (2413 BTU/hr) or 57 BTU per minute. Determine BTUs needed for a given temperature rise by the following relationship

$$\text{BTU} = (\text{wt})(\text{sp. heat})\ (\Delta T)$$

Divide the time estimate by 57×0.8 to arrive at heating time. If foods are frozen, consider 144 BTU per pound of latent heat of fusion—approximately 3 minutes additional heating for 1 pound product in a 1000-watt oven.

Appendix 2

Penetration of 2450 MHz microwave in various foods at 40° C (3,24)

Food	Penetration Depth, mm
Water, tap	15.00
Raw beef	10.50
Gravy	7.50
Pineapple	14.00
Cooked peas	12.00
Cooked cod	11.00
Cooked ham	5.00
Peanut butter	9.50
Liver pate	8.00
Macaroni and cheese	7.50
Ham salad	5.00
Mashed potato	7.00
Condensed gravy	6.00
Broth	8.00

Appendix 3

Specific Heat of Various Foods

	BTU/pound/°F
Water	1.00
High-water foods	
beverage	
cream	0.75
whole milk	0.92
skim milk	0.95
egg white	0.92
raw vegetables	0.95
fresh fruits	0.95
lean beef	0.82
lean pork	0.73
applesauce	0.96
fish	0.86
soup	0.94
cottage cheese	0.78
ice cream mix	0.80
Intermediate-moisture foods	
bread	0.68
dough	0.52
Fat and high-fat foods	
butter	0.50
margarine	0.45
vegetable oil	0.40
bacon	0.48
egg yolk	0.67
Dry foods	
sugar	0.30
raisins	0.47
flour	0.45
dried salted fish	0.42
dried vegetables	0.44
dried fruits	0.55
Air	0.24

Compilation by F.R.I. Enterprises

Appendix 4

Dielectric Properties of Foods at 2450 to 3000 MHz and 25 to 50°C (2,21,22,25,29)

	K′	K″	tan
Distilled water	83	11.3	0.13
Fruits, vegetables (dried)	10.5	5.5	0.52
Raw potato	65.0	22.0	0.33
Orange juice concentrate	54.0	15.0	0.27
Raw carrot	65.0	15.0	0.23
Raw apple	57.0	12.3	0.21
Peach	68.0	12.5	0.18
Pear	64.0	13.6	0.21
Egg white	69.7	15.8	0.22
Egg yolk	40.0	9.5	0.23
Mashed potato	65.0	22.1	0.34
Peas	63.0	15.7	0.25
Pineapple syrup	67.8	11.6	0.17
Frankfurters	39.0	26.9	0.69
Ham	66.6	47.0	0.70
Ice	3.2	0.0009	0.0029
Peanut butter	3.1	4.10	1.32
Corn kernel (15% moisture)	2.27	0.20	0.08
Whole wheat flour (14.7 moisture)	3.00	0.22	0.07
Pizza, baked dough	4.60	0.60	0.13
Pizza stuffing	10.10	3.10	0.30
French fried potato	12.4	4.60	0.37
Corn oil	2.62	0.17	0.06
Pork chop	49.8	18.3	0.37
Turbot fillet	53.6	14.1	0.26
Macaroni and cheese	54.5	20.7	0.38
Raw beef	46.0	12.6	0.27
Raw pork	67.0	41.0	0.61
Raw ham	69.3	43.9	0.63
Vegetable soup	65.8	16.9	0.26
Meat broth	73.9	21.4	0.29
Beef liver	38.9	17.2	0.44
Beef steak	48.9	18.0	0.37
Chicken	52.5	17.7	0.34
Salmon	49.8	17.2	0.29

Appendix 5

Useful Data and Constants (5,8,46)

Speed of light	3×10^{10} cm/sec
Relative permittivity	1.00
Dielectric constant, free space	8.854×10^{-12} F/m
Dielectric constant of water	80.000
Dielectric constant of bound water	4.50
Permeability of free space	1.257×10^{-6} Henry/m
Relaxation time of water	0.25×10^{-10} seconds
Relaxation time of food proteins	10^{-6} seconds
Intrinsic impedance	
air	377
water	42
0.1 M NaCl	42
food solids	188
potato solids	200
raw potato	50
moist potato solids	460
olive oil	218
Avogadro's number	$= 6.0225 \times 10^{26}$k mole $^{-1}$
Boltzmann's constant	$= 1.3805 \times 10^{-23}$ J/K
Electrical dipole moment	$= 8.6171 \times 10^{-5}$ ev/K
Electronic charge	$= 3.335 \times$ 10–30 cm
Electron rest mass	= 1 debye
Electron volt	$= 1.6021 \times 10^{-19}$ C
Planck's constant	$= 9.1091 \times 10^{-31}$ kg
	= 23.04 k cal/mole
	= 96.4 kJ/mole
	$= 6.6256 \times 10^{-34}$ Js

Appendix 6

Important Physical Relationships and Definitions

1. Energy = (Planck's constant) (Frequency)

$$= (6.626 \times 10^{-27} \text{erg.second}) (2450 \times 10^{6} \text{ Hertz})$$

$$= 4.51 \times 10^{-34} \text{ watt. hour} \approx 10^{-5} \text{ electron volt.}$$

2. Complex Permittivity, K*

$$= K' - jK''$$

where K′ is dielectric constant,
K″ is loss factor, and
j (the complex number) = $\sqrt{-1}$

3. Loss Tangent, tan δ

$$= \frac{K''}{K'}$$

4. Intrinsic Impedance, Z

$$= \frac{\text{(Electric field)}}{\text{(Magnetic field)}}$$

$$= \frac{\sqrt{K*}}{\mu}$$

where μ is permeability.

5. Attenuation Factor, α

$$= \frac{2\pi}{\lambda_o} \sqrt{\frac{K'[(1 + \tan^2\delta)^{1/2} - 1]}{2}}$$

$$= \frac{2\pi f}{3 \times 10^{10}} \sqrt{\left[\frac{K'}{2}\left[1 + \left(\frac{K''}{K'}\right)^2\right] - 1\right]^{1/2}}$$

where f is frequency, and

λ_o is wavelength in space.

6. Penetration Depth, D

$$= \frac{1}{\alpha} = \frac{\lambda_o}{2\pi} \frac{2}{\sqrt{K'\left[(1 + \tan^2\delta)^{1/2} - 1\right]}}$$

$$D_{50} = \frac{0.189\lambda_o}{\sqrt{K'}\sqrt{(1 + \tan^2\delta) - 1}}$$

$$\text{and } D_e = \frac{\lambda_o\sqrt{K'}}{2\pi K''}$$

7. Power absorbed at distance x, P_x

$$= P_o e^{-2\alpha x}$$

$$\text{and } X = \frac{1}{2\alpha} \ln \frac{P_o}{P_x}$$

8. Power absorbed per unit volume, P in watts/cm^2

$$= kfE^2K''$$

where k is a proportionality constant, 55.6×10^{-4}, in cgs system,

f is in Hz,

electric field E in volts/cm, and

K″ in Farad/cm.

9. Rate of heating

$$\frac{\text{Temp. Rise}}{\text{Time}} = \frac{dT}{dt} = \frac{14.32P}{\rho\, C_p}$$

where P is power density of foods,

ρ is density of food in gm/cm^3, and

C_p is specific heat.

10. Relaxation Time, τ

$$= \frac{1}{2\pi f_c} = V\frac{3\eta}{kT}$$

where f_c is critical frequency of maximum absorption,
V is the volume of a dipole,
η is viscosity of foods, and
k is Boltzmann's constant.

11. Electrical conductivity, σ'

$$= \omega K_o K''_\sigma \text{ or}$$

$$K''_\sigma = \frac{\sigma}{\omega K_o} \text{ and}$$

$$\sigma' = \frac{AC}{1000}$$

where K_o is dielectric constant in vacuum,
K''_σ is ionic loss factor, and
AC is equivalent conductivity.

12. Wavelength in foods (medium), λ_m

$$= \frac{\lambda_o}{\sqrt{K'}}$$

Appendix 7

Dielectric Properties of Water vs Temperature

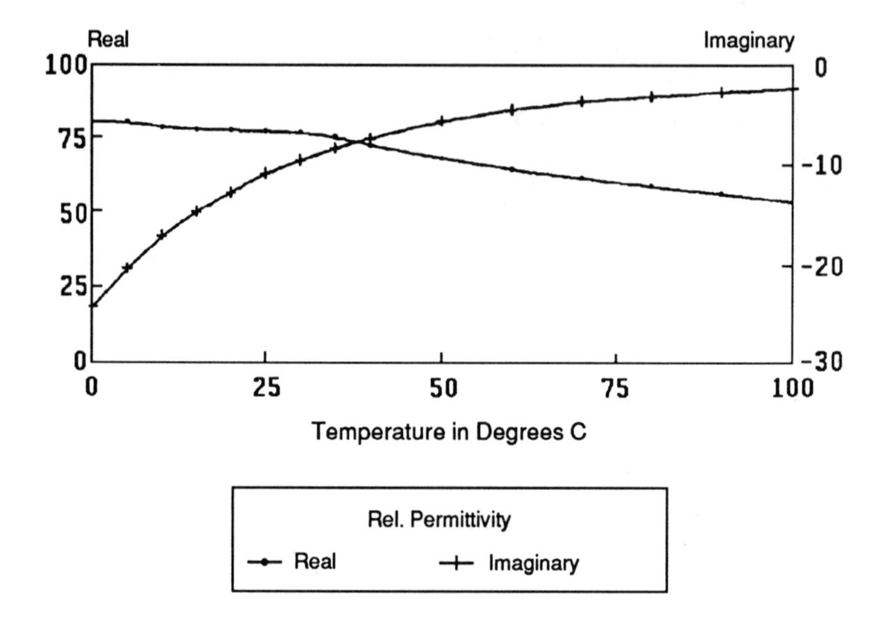

Figure 15-16. Dielectric properties of water vs. temperature. (Source: NBS Circular 589)

References

1. Bengtsson, N.E., and Ohlsoon, T. "Application of microwave and high frequency heating in food processing." Food Process Eng. Second Intl. Congress on Engineering and Foods. 8th European Food Symposium, Finland. 1979.
2. Bengtsson, N.E., and Risman, P.O. "Dielectric properties of foods at 3 GHz as determined by a cavity perturbation technique. II. Measurements on food materials." *J. Microwave Power* 6(2):107–124. 1971.
3. Bookwalter, G.N., Shukla, T.P., and Kwolek, W.F. "Microwave processing to destroy Salmonellae in corn-soy-milk blends and effect on product quality." *J. Food Sci.* 47:1683. 1983.
4. Benzason, A., and Edgar, R. "Microwave tempering in the food processing industry." *Electron Progress* Vol. XVII (1):8–12. Rathyon Co. 1976.
5. Condn, A.U., and Odisha, H. *Handbook of Physics*, Chapter 7, McGraw Hill, New York. 1958.
6. Cone, M., and Snyder, T. *Mastering Microwave Cooking*, Simon and Schuster, New York. 1986.
7. Copson, D.A. *Microwave Heating*, AVI/Van Nostrand Reinhold, New York. 1975.
8. Decareau, R.V. "Microwave in food processing." *Food Technology in Australia* 35:18. 1983.
9. Decareau, R.V. "Microwave in food processing." *Food Technology in Australia* 36:81. 1984.
10. Decareau, R.V. *Microwave in the Food Processing Industry*. Academic Press, New York. 1984.
11. Edgar, R. "The economics of microwave processing in the food industry." *Food Technology*, June, p. 106. 1986.
12. Federal Register. Food Additive, Subchapter 6 - Radiation and radiation sources intended for use in the production, processing, and handling of foods. Revised paragraph 121.3008, food additives regulation. Part 121, vol. 23(45), March 6, 1968.
13. Frohlich, H. *Theory of Dielectrics*, Oxford University Press, Ely House, London. 1958.
14. Gerling, John E. "Microwave oven power: A technical review." *J. Microwave Power* E2 22:199. 1987.
15. Hasted, J.B. "The Dielectric Properties of Water." *Progress in Dielectrics* 3:101–150. Haywood and Co. Ltd., London. 1961.
16. Jain, R.C., and Voss, W.A.G. "Dielectric measurements of browning agents, additives, and other high loss materials." *J. Microwave Power* E2 21(2):120. 1986.
17. Kenyon, E.M. "The feasibility of continuous heat sterilization of food products using microwave power." U.S. Army Natick Laboratory Tech. Report 71-8-FL (AD-715853). 1970.
18. Meisel, N. "Microwave applications to food processing and food systems in Europe." *J. Microwave Power* 8(1):143–148. 1973.

19. Misra, D.K. "Permeability measurement of modified infinite samples by a directional coupler and sliding load." *IEEE Transactions on Microwave Theory and Techniques* 29(1):65. 1981.
20. Moore, N.H. "Microwave energy in the food field. 20th Anniversary Meeting of R & D Associates." *Misc. Eng. Applications Newsletter* 1(1):5–7. 1968.
21. Mudgett, R.E., Goldblith, S.A., Wang, D.L.C., and Westphal, W.B. "Prediction of dielectric properties in solid foods of high moisture content at ultrahigh and microwave frequencies." *J. Food Process and Preservation* 1:119. 1971.
22. Mudgett, R.E. "Electrical properties of foods in microwave processing." *Food Technology* 36(No.2):104. 1982.
23. Mudgett, R.E., and Schwartzberg, M.G. "Microwave food processing—pasteurization and sterilization: a review." *Food Process Engineering* 78:1–11. 1982.
24. Mudgett, R.E. "Microwave properties and heating characteristics of food." *Food Technology* 40(No. 6):84-93,98. 1986.
25. Ohlsoon, T., and Bengtsson, N.E. "Dielectric food data for microwave sterilization processing." *J. Microwave Power* 10(1):93–108. 1973.
26. Osepchuk, J.M. "A history of microwave heating applications." *IEEE Transactions on Microwave Theory and Techniques* 32(9):1200–1224. 1984.
27. Pei, D.C.T. "Microwave baking—new developments." *Baker's Digest.* (February 1982):8–10, 32–33.
28. Pressman, T. *The Great Microwave Dessert Cookbook.* Contemporary Books, Inc., Chicago, IL. 1985.
29. Richman, P.O., and Bengtsson, N.E. "Dielectric properties of foods at 3 GHz as determined by cavity perturbation technique—measurement of food materials." *J. Microwave Power* 6(2)101–124. 1970.
30. Rothymayer, W.W. "Food Process Engineering." *Proceedings Royal Soc.*, London B191:71–78. 1975.
31. Roebuck, B.D., and Goldblith, S.A. "Dielectric properties of carbohydrate-water mixtures at microwave frequencies." *J. Food Science* 37:199. 1972.
32. Rosenberg, U. and Bogl, W. "Microwave thawing, drying, and baking in food industry." *Food Technology* 41(No. 6):85-91. 1987.
33. Rosenberg, U. and Bogl, W. "Microwave pasteurization, sterilization, blanching, and pest control in the food industry." *Food Technology* 41(No.6):92–99. 1987.
34. Schiffman, R. et al. "Application of microwave energy to doughnut production." *Food Technology* 25(No. 6):718–722. 1971.
35. Schiffman, R. "The application of microwave to the food industry in the United States." *J. Microwave Power* 8(2):137–142. 1973.
36. Schiffman, R. "Food product development for microwave processing." *Food Technology.* 40(No. 6):54–98. 1986.

37. Shaw, T.N., and Galvin, J.A.. "High frequency heating characteristics of vegetable tissues determined from electrical conductivity measurements." *Proceeding REE* 37:83. 1949
38. Schmidt, W. "The heating of food in a microwave cooker." *Phillips Tech. Rev.* 22(3):89–101. 1961.
39. Shukla, T.P. and Misra, D.K. Unpublished data. 1989.
40. Shukla, T.P. "Microwavable foods: dielectric properties of ingredients." Intl. Conference on Microwavable Foods, Schothland Business Research Inc., Princeton, NJ. 1988.
41. Shukla, T.P. "Hydrocolloids for microwavable foods." Intl. Conference on Formulating Foods for the Microwave Oven. The Packaging Group, Inc., Milton, NJ. 1989.
42. Singh, S.P., and Misra, D.K. "A magic T and sliding short technique for measuring the permittivity of modified infinite samples." *AEU* 35(3):137. 1981.
43. Smith, F.J. "Microwave hot air drying of pasta, onions, and bacon." *Microwave Eng. Applications Newsletter* 12(6):6–12. 1979.
44. Suzuki, T., Balwin, R.E., and Korsdigen, B.M. "Sensory properties of dextrose and sucrose cured bacon: microwave and conventionally cooked." *J. Microwave Power* 19(3):9. 1984.
45. Tinga, W.R. "Multiphase dielectric theory—applied to cellulose mixtures." Ph.D. Diss., Dept. of Electrical Engineering, University of Alberta, Edmonton, Alberta, Canada. 1969.
46. VonHippel, A.R. *Dielectric materials and Applications*, MIT Press, Cambridge, MA. 1954.
47. Voss, W.A.G. "Advances in use of microwave power." HEW Seminar, April 25, paper 008, p.2. 1969.
48. Vermeulen, F.E., and Chute, F.S. "The classification of processes using electric and magnetic fields to heat materials." *J. Microwave Power* E2 22(2):187. 1987.

16

Nonfood Uses for Sucrose

Charles B. Broeg *

"Why should common sugar, sucrose, be of interest for nonfood applications?" Among the many reasons to consider sucrose as a nonfood raw material are: (a) it is a continuously available, renewable resource; (b) it is produced in large quantities annually (approximately 100 million metric tons [1]); (c) it is available routinely as a high-purity material (less than 0.1 percent of impurities); and (d) it has several functionally useful chemical and physical properties.

There are several ways in which sucrose is used or has the potential for use in nonfood applications. Among these are as a raw material for manufacturing organic chemicals, in fermentation processes, and for pharmaceutical uses such as in tableting. These and other nonfood uses will be discussed in this chapter.

Most of the annual production of sucrose is used as food or in the manufacture of food. However, substantial quantities are usually available on a demand basis, and the infrastructure exists for increasing production by significant amounts provided adequate advance notice is given and suitable economic incentives exist. It is estimated that less than 0.5 percent of the annual production of sucrose is used currently for nonfood purposes.

Sucrose is a disaccharide containing both glucose and fructose molecules. Its major chemical properties are related to the three primary and five secondary hydroxyl groups present in the pyranose and furanose rings which are the major structural features of the sucrose molecule. In addition to Chapter 4 of this book, an excellent

*Charles B. Broeg, Consultant, Houma, LA, was formerly Vice President and Technical Director, Revere Sugar Corporation (Sucrest Corp.), Brooklyn, NY.

review of the chemical properties and reactivity of sucrose is given by Kollonitsch (2) and more recently by Khan (3).

Physical as well as chemical properties are important in the nonfood applications of sucrose. An older review of the physical properties may be found in Circular C 440 (4) published by the National Bureau of Standards. This circular presents tabulated information on optical rotation, solubility in water, density and viscosity of aqueous solutions, refractive indexes, and crystallography. Additionally, it is a source of analytical methods as well as methods for synthesizing many sugar derivatives. Continuing work on sucrose properties, especially those associated with analytical procedures, are carried on by members of the International Commission for Uniform Methods of Sugar Analysis. Results of these research efforts are published in their Proceedings (5) following each quadrennial meeting, the last of which occurred in 1986. Also, see Chapter 17 in this book.

The solubility of sucrose in water and other solvents is especially important when considering its potential as a raw material for the manufacture of numerous organic chemicals. Organic solvents facilitate the reaction of sucrose with organic chemicals because many of these reactants are not soluble in water, or react with water in preference to sucrose. Lack of information about the solubility of sucrose in organic solvents led the Sugar Research Foundation forty years ago to fund a project for the development of this kind of information. Herstein Laboratories, Inc., reviewed the scientific literature and studied solubility in many organic solvents to provide information greatly needed for the synthesis research on sucrose (6). An important tabulation of the physical properties of sugar solutions and solutions containing other sugars was published by the British Food Manufacturing Industries Research Association (7). The original project funded by the Sugar Research Foundation, and its successor organizations, has been responsible for much of the sugar research of the past four decades. Summaries of the various sizes as well as the types of sugar commercially available are found in the *Handbook of Sugars* by Pancoast and Junk (8) and in this book.

In addition to chemical synthetic reactions, there are other nonfood uses for sucrose. These fields are diverse and frequently depend on the physical rather than the chemical properties of sucrose. They include medicine and pharmaceutics, fermentation, and several nondescript applications that defy classification. Sucrose as a chemical raw material, however, is the field that offers greatest potential for this unusual, naturally occurring product.

Sucrochemistry

Chemical reactions of sucrose and other sugars have been intensively explored over a great many years. Much of the earlier work was undertaken as a tool to determine the structures of sugars in general (9,10,11). The systematic investigation of sucrose and by-products of the sugar industry as raw materials for the manufacture of chemical substances received a major thrust when the Sugar Research Foundation was organized in 1943 (2). The effort was continued as part of the International Sugar Research Foundation when the Sugar Research Foundation was reorganized to include participation of worldwide sugar companies (12). Activities were encouraged by the awarding of grants to researchers to explore sucrose as a chemical raw material.

The term "sucrochemistry" originated with H. B. Hass, who served as the scientific director of the Foundation for a number of years. Dr. Hass explained sucrochemistry to the American Chemical Society (ACS) at a 1976 symposium (13) jointly sponsored by the International Sugar Research Foundation and the Division of Carbohydrate Chemistry of the ACS.

These Foundations were very successful in stimulating thinking and research on the chemical reactions of sucrose and invert sugar (glucose and fructose), a by-product of sucrose from its ready hydrolysis under acidic conditions. Some of the early research works, although not successful by some standards, did educate researchers in the difficulties encountered with sucrose reactions and in establishing the conditions under which sucrose reactions could be successfully carried out.

The scope of the early work was summarized by Long in 1949 (14). This report reviews older literature and patents relating to the conversion of sucrose into various derivatives and into "plastics." Among the references are those referring to the production of dextrose (glucose) from sucrose. Substantial coverage was also given to invert sugar and other by-products of the manufacture of sugar.

An annotated review of the earlier patents on sugar reactions was prepared by Hunt (15) in 1961 under direction of the Sugar Research Foundation. This compilation also covered patent literature relating to other sugars and molasses. This review is especially useful because the patents were grouped according to the type of reaction involved. This annotated bibliography, and the bibliography in the Long report referred to above, are good guides to the early sucrochemistry scientific and patent literature.

The magnitude of the Foundation's research efforts on sugar warranted the publication of a 25-year review, summarizing the results of the various projects funded during this period (2). The projects were assessed with regard to their accomplishments and the potential commercial applications of the chemical derivatives. A result of these twenty-five years of effort was the demonstration that sucrose is a reactive molecule under the appropriate conditions and that it will undergo a wide range of chemical reactions.

Ethers

Several procedures for synthesizing sucrose ethers were developed. Older methods were based on reactions with organic sulfates and organic halogen derivatives to prepare methyl ethers. Newer methods include the preparation of sodium sucrates and their subsequent reaction with an organic halide to form the desired ether. The Kollonitsch review (2) importantly points out failures as well as successes and the importance of using the correct solvent for the reaction. Properties of several sucrose ethers and their potential applications as, among others, surfactants and surface coatings are included in the Kollonitsch publication.

Esters

A second major class of sucrose derivatives is esters. These include the mono- and polysubstituted esters of the usual aliphatic acids (acetic, propionic, etc.) and the longer-chain acids including saturated and unsaturated fatty acids. These esters, particularly the fatty acid esters, were investigated intensively because of their large commercial potential in the manufacture of detergents. The commercial success of some of these has been outlined by Kosaka and Yamada (16) and by Parker, James, and Hurford (17). Three of the most important procedures for synthesizing esters are:

a. Reacting sucrose with organic acid chlorides or acid anhydrides usually results in a mixture of mono- and polysubstituted esters. Pyridine is the most common solvent for acid chloride reactions. Monoesters can be obtained under carefully controlled conditions (2, pp. 52–53).

b. Transesterification of sucrose with the methyl ester of the desired acid was developed for economic reasons and for better control over the production of "pure" monoesters. This method can be used to produce polysubstituted esters as well as the monoesters. Various modifications have been developed to give the desired product (18). The most common solvent is dimethylformamide.

c. More recently, researchers at Tate and Lyle, Ltd., have developed a "solventless" process for making sucrose esters of several common fats. Fats melted at a high temperature become the solvent for sucrose. This reaction results in a mixture of products including sucrose esters, glycerol, and mono- and diglycerides. This mixture presents a formidable separation task. However, since a primary objective of this reaction is a product for detergent applications, separation is not necessary.

The commercial potential of the saturated and unsaturated fatty acid esters of sucrose has led to extensive investigation of the usefulness of these products (13,18). Special pilot plant studies have been undertaken to develop a commercial process for manufacturing "drying oil" esters for use in the ink, paint, and protective coating industries (13,19). Additional esters are found in chapter 4 of the Kollonitsch review (2). These include esters of sulfonic acids, sulfuric and nitric acids, carbonates, tricarbonates, acrylates, and others. The use of various esters has been reviewed by Kosaka and Yamada (13, chapter 5), Bobalek (13, chapter 12), Faulkner (13, chapter 13), Weaver (13, chapter 14), Coney (13, chapter 15), and Liva and Anderson (13, chapter 16). An update was presented by Hickson in 1978 (3, chapter 6).

Miscellaneous Reactions

Sucrose reacts with many chemicals in addition to those that have already been described. Among the more widely investigated reactions are the following:

a. *Aldehydes and ketones* Sucrose reacts to form cyclic acetals and ketals. If other alcohols are included with sucrose in the reaction mixture, mixed acetals are formed under suitable conditions (2, chapter 5).

b. *Alkali metals* Sucrose reacts with sodium, lithium, and potassium to produce mono- and polysucrates. These compounds are useful as intermediates in the manufacture of other sucrose derivatives. These metal derivatives and their use as intermediates are discussed by Kollonitsch (2, chapter 9).

c. *Oxidation* Sucrose is readily oxidized by several oxidizing agents to produce mixtures of acidic compounds (2, p. 106). Caramel is produced by controlled alkaline degradation.

d. *Pyrolysis* Pyrolysis of sucrose leads to a great many compounds and great amounts of carbon dioxide and carbon. Pyrolysis has not proven to be a suitable approach to the commercial production of other chemicals (21, 22).

e. *Hydrogenation* Catalytic hydrogenation ruptures the sucrose molecule and produces many hydrogenation products. The best known are sorbitol and mannitol. Severe hydrogenating conditions usually result in further degradation to yield various alcohols, glycols, and glycerol (2, chapter 7).

f. *Reductive aminolysis* Reaction of sucrose with ammonia, or organic amines under reducing conditions, results in a number of nitrogen-containing compounds. A major product of the reaction with ammonia is 2-methyl-piperazine (2, chapter 7).

g. *Halogenation* Sucrose reacts with chlorinating agents under appropriate conditions to produce chloroderivatives (hydroxyl groups replaced by chloride groups). Considerable interest in these compounds has developed because of the intense sweetness of one of them (2, chapter 8; 13, chapters 2 and 4; 3, chapter 7).

h. *Others* Sucrose undergoes reactions with epoxy compounds to form epoxides. These compounds and related derivatives are useful in the manufacture of rigid polyurethane foams (13, chapters 17, 18, and 19). In other reactions, sucrose may provide a means of modifying the properties of other compounds such as organo-tin pesticides (13, chapter 11). In others, the sucrose derivative may become part of a polymeric substance designed for specific applications (19,20).

References 2, 3, 12, and 13 contain bibliographies of more extensive literature on sucrose reactions and potential areas of use.

Fermentation Products

Sucrose is also used in fermentation processes. The oldest and best known fermentation product is ethanol. Sucrose is not an essential ingredient for making fermentation ethanol, but its presence in an inexpensive sugar industry by-product such as molasses, and the ease with which it can be converted into fermentable hexoses, have made it a useful raw material for several other fermentation processes.

Ethanol

More than fifty years ago, blackstrap molasses was used to produce substantial quantities of ethanol (23, Vol. I, chapter 3). However, chemical processes for converting petroleum raw materials to ethanol were developed, and fermentation ethanol was largely displaced by petroleum ethanol as an economic fact of life. The abrupt increases in the price of petroleum in the 1970s revived interest in fermentation ethanol, not as ethanol but as an octane-enhancing component of gasoline for automobiles. Despite high petroleum prices, it has been necessary to subsidize fuel ethanol by the abatement of taxes. Gasahol (24) has received a great amount of attention in recent years, but the most important source for fuel ethanol in the United States has been corn. The potential for manufacturing ethanol from sugar has been extensively investigated (25, 26), but its economics in the U.S. are not promising at this time. Brazil, on the other hand, has made the production of fuel ethanol from sugarcane a national policy (27).

Others

Ethanol has been the major fermentation chemical produced from sugar, but several other organic products have been produced by fermentation processes. Many of these are reviewed extensively in *Industrial Fermentation* (23). Manufacture of acetic, lactic, and citric acids as well as that of acetone and butanol are reviewed. Choice of fermentation raw material is usually dictated by a combination of factors such as suitability, availability, and cost.

By-product molasses from the sugar industry continues to be an important raw material for citric acid production (28). Dextran, a fermentation polymer produced from sucrose by *Leuconostoc mesenteroides*, is unique in that sucrose is essential for its production.

Dextran has long been a plague to the sugar industry. Its inadvertant presence in sugarcane juices results from a variety of conditions existing during the harvesting and subsequent processing of sugarcane. Formation of dextran not only reduces the amount of recoverable sucrose, but also gives rise to problems in the sugar recovery processes of refineries and raw sugar mills. However, this problem has been brought under control. The sugar industry has developed a process for converting dextran contaminants into products with commercial utility. Wentrzell reported its production as a by-product of sugar manufacturing in 1946 (29). This report was followed by one referring to dextran as a blood plasma extender in 1947 (30). Owens reported on its production in 1948 (31) and again on its potential applications (32) including its use as an additive for oil drilling fluids (33).

Dextran's use as a blood plasma substitute was reviewed by Harrison in 1954 (34). A book entitled *Dextran*, reporting its uses in medicine, was published in 1955 (35). Current medical uses of dextran products manufactured by Pharmacia Laboratories are described in the *Physician's Desk Reference* (36). Johnson and Johnson distributes a dextran molecular sieve for absorbing wound exudates. It is packaged under the registered trademark "Debrisan."

Pharmaceutical Applications

Sucrose has many applications which do not depend on its chemical reactivity. Many of these uses depend upon its physical properties. The pharmaceutical industry uses sucrose in several different forms in a variety of preparations.

Sucrose has found use in elixirs and syrups not only because of its sweet taste but also because of its solubility and the bodying effects it imparts to such products (37,38). Crystalline and finely pulverized sucrose is used in a number of pharmaceutical preparations as a diluent, excipient, or binder (37).

The manufacture of pharmaceutical tablets may incorporate sucrose as a diluent to control concentration of active ingredients, as a binder to hold ingredients together in a formed tablet, and, when prepared in suitable granulations, to cause powdery mixtures to flow readily in high-speed tablet presses (37, 38). Special sucrose products have been manufactured into "direct compression vehicles" for tablet production. When sucrose is pulverized or crystallized into particles of less than 30 microns in size, it adds a chewable character to tablets. Vitamin preparations are frequently formed as chewable tablets. Before being used for such purposes, it is necessary to agglomerate or granulate the fine sucrose into a size that makes it compatible with flow requirements for a direct compression vehicle (37, 38, 39).

Tablets and other pharmaceutical preparations are frequently coated to give a protective finish to the product. Coating methods are essentially those employed by the confectionery industry (40). A well-made sugar coating protects tablets from chipping and dusting and from the widely differing atmospheres the tablets might encounter.

Sucrose is also used as a diluent, excipient, or base for preparing encapsulated and sustained-release dosage pharmaceuticals (38). In one type of preparation, sucrose crystals are used as a base to form solid spheres upon which active ingredients can be deposited. In this type of preparation, interlayers of active ingredients and inactive materials can be alternately deposited upon the base sphere, making it possible to incorporate more than one active ingredient in the product and to control the sequence and time of release of the drug. This technique is another one that has been borrowed from the confectionery industry where the spherical base is known as a nonpareil. In practice, sugar crystals are coated in a revolving pan with alternate dosings of sugar syrup and dry starch. After the crystal has been converted into a sphere, it is built up to the desired size by further layerings of sugar syrup and starch.

The use of dextran as a blood plasma extender has been discussed on page 283 of this chapter.

Folk medicine has produced an unusual use of sugar for the treatment of wounds that resist conventional methods. The direct application of sugar to "incurable" wounds was reported by L. Herzage and J. R. Montenegro in an *Argentine Surgical Bulletin*, October, 1980 (27). Additional exploration of the use of sugar to treat resistant wounds has been reported in *Sugar y Azucar* (41). Hough (20) has reported on the use of the sodium and aluminum salts of sucrose octasulfate as an antipeptic and antiulcer agent.

Cosmetics Applications

Both mono- and polyesters of sucrose have been reported to have desirable properties for use in various types of cosmetics. Kollonitsch (2) reported the desirable properties of sucrose esters as emulsifiers. More recently, Desai and Lowicki (42) reviewed several sucrose surfactants and their desirable properties for use in cosmetics. Ester use in lotions was reviewed by Robinett in *Cosmetics and Perfumery* (43). Neulinger (44) has reported on the incorporation of sucrose esters in cosmetic sticks and the properties of such sticks when compared with standard sticks.

Other Applications

Sucrose has been used for a number of other purposes. Among these are uses as a binder in foundry sand molds, and in clay and other ceramics (45,46). In similar uses, it has been claimed as a reducing agent as well as a binder (47). An unusual combination of properties has been reported by Hutchison and Johnson, who used sucrose to control the density of thorium oxide fuel pellets where it also served as a binder (48). Other uses include those as an additive to electroplating solutions (49) and those for retarding the setting times of cement (50).

The various uses of sucrose reported here are but a few of the great number reported in both patent and scientific literature. However, there is still room for new ideas and new uses.

Conclusion

The preceding review summarizes the developments over the past forty years that have greatly broadened the potential of sucrose as a nonfood resource. The major change in the chemical processes utilizing sucrose has been the shift from reactions in aqueous solutions to those in nonaqueous solvents. This shift has led to many sucrose derivatives that could not be synthesized in a water solvent.

Nonfood uses are small compared with food uses. However, the food demand for high-purity sucrose assures the availability of this organic compound for other potential applications. Sucrose esters of fatty acids are probably the largest nonfood use of sucrose. Interestingly, a sucrose polyester used for a food application may become a significant outlet for sucrose. The Procter and Gamble Company has synthesized a polyester that is not utilized by human digestive processes. The product has many of the attributes of a fat, and studies are underway to utilize the product as a noncaloric fat substitute.

Future developments in nonfood uses of sucrose will probably arise from the application of biochemical procedures and enzymes for synthesizing a different group of sucrose end-products.

References

1. *Sugar and Sweetener Situation and Outlook Report Yearbook*, (SSR 14 N 2), Table 1, page 35; Economic Research Service, U.S. Department of Agriculture, Washington, D.C. 1989.
2. Kollonitsch, V. *Sucrose Chemicals.* International Sugar Research Foundation. 1970.
3. Khan, R. *Sugar: Science and Technology*, pp. 181–210. Edited by Birch, G. G., and Parker, K. J. Applied Science Publishers, London. 1978.
4. Bates, F. J., and Associates. *Circular of the National Bureau of Standards C 440.* U. S. Government Printing Office, Washington, DC. 1942.
5. *Report of the Proceedings of the Nineteenth Session.* International Commission for Uniform Methods of Sugar Analysis, Peterborough, England. 1986.
6. Kokonenko, O. K., and Herstein, K. M. "Non-Aqueous Solvents for Sucrose." *Chemical and Engineering Data Service* 1(1):87. 1956.
7. Norrish, R. S. *Selected Tables of Physical Properties of Sugar Solutions*, No. 51, British Food Manufacturing Research Assoc., Leatherhead, England. 1967.
8. Pancoast, H. J., and Junk, W. R. *Handbook of Sugars*, 2d ed., AVI/Van Nostrand Reinhold, New York. 1980.
9. Hudson, C. S., and Johnson, J. M. *J. Am. Chem. Soc.*, 37:2748. 1915.
10. Haworth, W. N. *J. Chem. Soc.*, 107, No. 8:8. 1915.
11. Fischer, E., and Oetker, R. *Ber. Deut. Chem. Ges.*, 46:4029. 1913.
12. Cheek, D. W. *Sugar Research 1943–1972.* Int. Sugar Research Foundation. 1974
13. *Sucrochemistry* Edited by Hickson, J. H. Am. Chem. Soc. Symposium, Washington, DC. 1977.
14. Long, L., Jr. *Sugar and Sugar By-Products in the Plastics Industry.* Technological Report Series #5, Sugar Research Foundation, New York. 1949.
15. Hunt, M. *Patents on the Reactions of Sugars.* Sugar Research Foundation. New York. 1961.
16. Kosaka, T., and Yamada, T. *Sucrochemistry*, Edited by Hickson, J. H. Am. Chem. Soc. Symposium, Washington, D.C. 1977. pp. 84–96.
17. Parker, K. J., James, K., and Hurford, J. *Sucrochemistry*, Edited by Hickson, J. H. Am. Chem. Soc. Symposium, Washington, D.C. 1977. pp. 97–114.
18. *Sucrose Ester Surfactants.* Research report, Sugar Research Foundation, New York. 1961.
19. *Engineering and Pilot Plant Data for the Commercial Production of Sucrose Esters for the Ink, Paint, and Protective Coating Industries.* Sugar Research Foundation, New York. 1963.
20. *Expansion of Sugar Uses Through Research.* International Sugar Research Foundation, Bethesda. 1972.
21. Johnson, R. R., Alford, E. D., and Kenzer, G. W. *J. Agr. Food Chem.*, 17(1):22. 1969.

22. Fagerson, I. S. *J. Agr. Food Chem.* 17(4):747-750. 1969.
23. *Industrial Fermentation*, Edited by Underkoffler, L. A., and Heckey, R. J. Chemical Publishing, New York. 1954.
24. Hunt, V. *The Gasahol Handbook*. Industrial Press, New York. 1981.
25. Lipinski, et al. *System Study of Fuels from Sugar Cane, Sugar Beets, and Sweet Sorghum*. Battelle Columbus Laboratories, Columbus, OH. 1976.
26. *The Report of the Alcohol Fuels Policy Review*. U. S. Dept. of Energy, Washington. 1979.
27. de Almeida, H. "Brazil's Sugar and Alcohol Programmes." World Sugar Research Organization Symposium Report, Buenos Aires. 1981.
28. Noyes, R. *Citric Acid Production Processes*. Noyes Development Corp., Park Ridge. 1969. *Beets, and Sweet Sorghum*. Battelle Columbus Laboratories, Columbus, OH. 1976.
29. Wentrzell, T. *Socker Handl.* 2:263–80. 1946.
30. Ingelman, B. *Act. Chim. Scand.* 1:731–8. 1947.
31. Owens, W. L. *Sugar* 42(8):28–9. 1948.
32. Owens, W. L. *Sugar* 45(3):42–3. 1950.
33. Owens, W. L. *Sugar* 45(11):35–7. 1950.
34. Harrison, J. H. *Ann. Surg.* 139:137–42. 1954.
35. *Dextran*, C. C. Thomas, Springfield, IL. 1955.
36. *Physicians' Desk Reference*, 39, Medical Economics Co., Oradell, NJ. 1985.
37. *U. S. Pharmacopeia*, 19, 20. U. S. Pharmacopeial Convention, Rockville, MD. 1975, 1980.
38. *The Theory and Practice of Industrial Pharmacy*. Edited by Lachman, L., Lieberman, H. A., Kanig, J. L. Lea and Febiger, Philadelphia. 1970.
39. Mendes, R. W., Gupta, M. R., Katz, I. A., and O'Neil, J. L. "Nu-Tab as a Chewable Direct Compression Carrier." *Drug and Cosmetic*. December, 1974.
40. NCA/AACT Technical Seminar, 1987, *Manufacturing Confectioner*, November 1987, pp. 39–58.
41. Dyer, D. L. "Facts about Sugar." *Sugar y Azucar* 81(2):14–16. 1986.
42. Desai, N. B., and Lowicki, N. "New Sucrose Esters and Their Application in Cosmetics." *Cosmetics and Toiletries* 100(6):55–9. 1985.
43. Robinette, H., Jr., *Cosmetics and Perfumery* 89(3):63–4. 1974.
44. Neulinger, K. F., *Cosmetics and Toiletries* 92(7):65–7. 1977.
45. Starr, C. British Pat. 1,171,963, November 3, 1969.
46. Cooper, R. H. U. S. Pat. 3,307,959, March 7, 1967.
47. Adler, A. U. S. Pat. 3,285,734, November 15, 1966.
48. Hutchison, C. R., and Johnson, R. G. R. *WAPD TM 576*, Atomic Energy Commission, 1966.
49. Ababi, V., and Onee, P. Pat. RO 65,316., September 30, 1978.
50. Lange, B.A., Decker, H.C. U.S. Patent 4,375,987.

17

Methods of Analysis of White Sugar

S. E. Bichsel and Thomas Wilson *

On the following pages are methods of analysis for white sugars approved by the United States National Committee on Sugar Analysis.

*S. E. Bichsel is Senior Vice President Research and Development, Holly Sugar Corporation, a subsidiary of Imperial Holly Corporation, Colorado Springs, CO. Thomas Wilson is Refinery Chemist, Colonial Sugar Company, Gramercy, LA.

Method Number: USNC MOA 0001

Analysis: Color & Turbidity

Purpose: To determine the color and turbidity of white sugar.

Apparatus: Spectrophotometer with 420-nm and 720-nm wavelengths.
5-cm or 10-cm spectrophotometric cell.
Vacuum or sonic bath (to remove entrained air).

Reagents: Filtered, distilled, or deionized water.

Procedure: Prepare a 50% ± 0.2% solids solution of the sugar to be tested. Remove the entrained air via vacuum or sonic bath. Carefully pour the solution into the spectrophotometric cell and determine the attenuancy [A_c or -log (t)] of the solution at 420 nm and 720 nm, using filtered, distilled water as a reference standard.

Calculation: Color index (compensated for turbidity)—

$$\frac{1000\,[(A_c)_{420} - 2(A_c)_{720}]}{bc}$$

where: b = cell length (cm)
c = concentration (g/ml)
A_c = attenuancy = $-\log T_s$ = $2-\log \%T_s$

which reduces to: $163[(A_c)_{420}-2(A_c)_{720}]$ for a solution of 50% solids and cell length of 10 cm.

Turbidity = $\%T_s$ at 720 nm = $326(A_{c720})$

Alternate procedure: Same as above except include membrane filtration (1.2-micron filter) prior to air removal and omit reading at 720 nm. (Reading portion of sample before and after filtration allows turbidity by difference.)

Method Number: USNC MOA 0002

Analysis: Moisture

Purpose: Determination of the percentage by weight moisture in white sugar.

Apparatus: An oven of the following type:

1. Vacuum oven operated at 20 to 25 inches mercury at 80 to 85°C.
2. Forced-draft oven operated at 105°C.
3. Convection oven operated at 105°C.

Analytical balance (sensitive to 0.1 mg).
25-mm x 50-mm aluminum dishes with covers.
Dessicator with moisture-indicating dessicant.
Tongs to handle dishes.

Reagents: None.

Procedure: Dishes should be washed and dried with a clean towel and placed in an air oven to dry to 105°C for at least 30 minutes before use. Thirty minutes before use, remove dishes from oven and place in dessicator to cool to ambient lab temperature before determining tare weights. Weigh each dish and record the weight as W_1. From a representative well-mixed sample, approximately 10 g of sugar are added rapidly and the combined container and sample weight is determined (W_2). The container and sample are placed in the oven for three (3) hours. After the sample has dried, remove the container and sample from the oven and transfer to the dessicator to cool.

After cooling 20 to 30 minutes (to ambient), proceed to weigh the container and sample and record the weight (W_3).

Calculation: The loss in weight is expressed as a percentage of the original sample weight.

$$\% \text{ moisture} = \frac{100(W_2 - W_3)}{W_2 - W_1}$$

where: W_1 = weight of empty dish
W_2 = weight of dish and sugar before drying
W_3 = weight of dish and sugar after drying.

Method Number: USNC MOA 0003

Analysis: Grist

Purpose: To determine the percentage by weight of the various sugar granulations in the sample.

Apparatus: A nest of sieves (U.S. or Tyler) with pan and lid.
Ro-Tap sieve shaker or other appropriate type shaker.
Balance sensitive to 0.1 g.
Sample splitter.
Fine-bristle brush.

Reagents: None.

Procedure: Divide the sample in the sample splitter sufficient times to quarter it to slightly over 100 g. Weigh out 100 g and place on top screen of screen nest equipped with pan and lid. Place nest in sieve shaker and tap for 5 minutes. After tapping, put the retained sugar from each screen on a smooth paper, invert the screen, and brush the bottom with the fine-bristle brush. Weigh the fraction from each screen in the sieve nest and record the weight as the percentage retained on that screen.

Calculation: Record the weight retained on each sieve as the percentage retained.

Method Number: USNC MOA 0004

Analysis: Sediment

Purpose: Sediment in white sugar is determined by visually comparing a sediment test pad with standard sediment discs.

Apparatus: Filtering apparatus.
Vacuum source.
Appropriate filter paper (Whatman No. 40 or No. 54; E.D. 613 or 8613, Sediment Testing Supply TC 900).
Drying oven or heating lamp.

Reagents: Distilled water.

Procedure: From a representative sample, weigh out 300 g of sugar and dissolve in approximately 400 ml of hot distilled water. Place filter pad in filtration apparatus and apply vacuum. Pour sample into filter apparatus and continue filtration until all sample has been filtered. Rinse the sample beaker several times to ensure transfer of any sediment and finally rinse walls of sample apparatus to wash any sediment onto pad. Remove the pad from the filtering apparatus and place on a clean, flat surface. Dry the pad in the oven or under the heat lamp and compare to the standard sediment pads. Repeat the procedure with black sediment pads to determine colorless and white sediment.

Method Number: USNC MOA 0005

Analysis: Ash

Purpose: To determine the quantity of ash in white sugar.

Apparatus: Ash bridge or conductivity meter.
200-ml volumetric flask.

Reagents: Deionized water.

Procedure: Weigh out 50.0 g of sample and transfer to 200-ml volumetric flask. Add deionized water to dissolve the sugar and fill flask to near the mark. Attemperate to 20°C and make up to the mark. Rinse the cell and the cell vessel with several portions of deionized water, then rinse with the sugar solution and read the specific conductance.

Calculation: $L_{sugar} = L_s - 0.43(L_w)$

$\% \text{ Ash} = L_{sugar} \times 0.00055$

where: L_{sugar} = specific conductance due to sugar ash
L_s = specific conductance of sugar solution
L_w = specific conductance of water solution.

Method Number: USNC MOA 0006

Analysis: Sulfite (SO_2)

Purpose: To determine the quantity of SO_2 in white sugar.

Apparatus: 500-ml Erlenmeyer flasks.
1 burette, preferrably 10-ml capacity graduated to 0.05 ml for iodine titrations on sugar.
2 burettes, 50 ml, for standardization of solutions.

Reagents: 1 N potassium hydroxide solution.
Sulfuric acid-water, 1:3.
De Koninck's starch indicator solution.
Standard potassium dichromate solution (1 ml = 0.0002 g SO_2).
Standardized sodium thiosulfate solution.
Standardized iodine test solution.

Procedure: Dissolve 100 g of sugar in 150 ml of cold distilled water in a 500-ml Erlenmeyer flask. Add 10 ml of De Koninck's starch indicator solution and 25 ml of 1 N potassium hydroxide solution. Mix by gentle shaking. Allow to stand for a few minutes. Add 10 ml of 1:3 sulfuric acid, mix, and then without delay titrate with the standardized iodine test solution (1 ml = 0.0002 g SO_2, or equivalent as determined by standardization) until obtaining a faint blue color that remains for one minute.
Make a blank determination on water and reagents, using the same volume of water and reagents as used for the test on sugar. Deduct the amount of the blank from the titer of the sugar solution and calculate the total SO_2 expressed as parts-per-million of sugar from the corrected titer.

Calculation: (ml of iodine solution used for sugar solution – ml of iodine solution required for blank) × grams SO_2 equivalent to 1 ml of iodine solution × 10,000 = SO_2 in parts-per-million of sugar.

Notes: Keep all equipment clean.
The iodine test solution should be kept in a glass-stoppered, dark bottle and stored away from light. Any iodine solution as weak as this tends to lose its strength

imperceptibly. Hence, the solution should be rechecked at least once per week against the thiosulphate standard. The solution should not be allowed to come in contact with rubber tubing or rubber stoppers.

Method Number: USNC MOA 0007

Analysis: Floc

Purpose: To determine the extent of floc formation of an acidified white sugar solution.

Apparatus: Source of light giving strong pencil-like beam.
Clear glass jars of 1000-ml capacity.

Reagents: Orthophosphoric acid: 75g H_3PO_4/100 g (1.58 g/cm^3).

Procedure: Preparation of acidified sugar solution: 640 g sugar, 533 ml distilled water, and 3.5 ml of phosphoric acid are mixed with a stirrer in a glass jar until the sugar is dissolved. The jar containing the acidified sugar solution is set aside and observed on the third, seventh, and tenth days for floc formation. A blank is run on the water with each set of tests: 1000 ml of the water used to dissolve the sugar is acidified with 3.5 ml phosphoric acid, set aside, and observed at the same time as the test solutions are observed. If floc is observed in the blank, then the water must be treated to render it floc-free before repeating the tests.

Observation of floc: Care should be taken when moving the jars for observation. The solutions should not be agitated in any way as the floc, if present, is very fragile. The jar is placed in front of a strong pencil-like beam of light. The observer views the solution from the front of the jar, observing that part of the solution lighted by the beam of light. The bottom, top, and middle of the solution are observed. Floc may either rise, be suspended, or precipitate. All three of these conditions may exist in a single sample.

Calculation: After the tenth day of observation, a number is assigned to the sugar to express its quality. The *size* of the floc particles and not their quantity determines the rank, as indicated below.

Rating System for Floc:

0—negative Complete absence of visible particulate matter.

0—turbid Cloudy, but contains no visibly discrete particles.

1—pinpoint	Very small, discrete particles, the shapes of which are not discernible but which are visible in a strong beam of light.
2—light	Several particles gathered together to form a small, fleecy particle visible in a strong beam of light (approximate size 0.8 mm).
3—medium	A feathery-like particle visible in a strong beam of light (approximate size 1.5 mm).
4—heavy	An agglomerate of colloidal particles forming a large, fluffy particle, visible without the need for a strong beam of light (approximate size 3 mm).

Method Number: USNC MOA 0008

Analysis: Saponin

Purpose: To determine the saponin content of beet white sugar and thus predict the floc-forming potential.

Apparatus: Volumetric pipettes, 10- and 25-ml capacity.
Millipore filter holder, glass.
Millipore filters, white, 1.2-micron porosity and 47-mm diameter.
Micro pipettes, to deliver very small drops.

Reagents: Antimony pentachloride reagent ($SbCl_5$), 5% solution.
5 ml $SbCl_5$, 95 ml $CHCl_3$, and 1 g $SbCl_3$.
pH 2.0 HCl solution.
Control sugar (2 floc rating at the end of 72 hours).

Procedure:

1. Dissolve 100 g refined beet sugar in 200 ml of HCl solution. (Time 1 to 3 min)
2. Filter with vacuum through Millipore filter. (Time 1 to 2 min)
3. Rinse filtrate with 100 ml of HCl solution. (Time 1 min)
4. Oven dry Millipore filter at 150°C. (Time 2 to 5 min)
5. Apply a drop of 5% $SbCl_5$ reagent to Millipore filter and to the Control Millipore filter. Compare at room temperature and/or warmer temperatures.

Although high relative humidity may impede the saponin-antimony pentachloride pink color development at room temperature, this interference does not occur at temperatures as high as 105°C.
If not satisfied with the color development, then apply reagent again to the unused areas of the control and unknown filters for reevaluation. Millipore filters retained enclosed for as long as four months still give a positive saponin test.

Calculation: To estimate the floc-forming tendency, a section of the unused comparative limit standard should be placed side by side with the filter from the sugar under test. The antimony pentachloride reagent should be placed on the two filter sections in near-simultaneous fashion so that the color development interval is equal for each.

Judge the sugar to pass for soft drink production if the developed intensity of the spot on the sample disk is less than or equal to the spot intensity developed on the limit standard.

Reject and divert from soft drink production any sugar where the saponin intensity on the sample exceeds that on the limit standard.

Method Number: USNC MOA 0009

Analysis: Polarization

Purpose: To determine the polarization of white sugar.

Apparatus: Saccharimeter calibrated with quartz plates.
Flasks (100 ml) conforming to ICUMSA class A, which will be individually calibrated by weighing with water at 20.0 ± 0.1°C. Flasks whose contents fall within the range of 100.00 ± 0.01 ml may be used without correction. Flasks whose contents fall outside this range may be used with the appropriate correction to 100.00.
Polarimeter tube, length 200 mm (or of length specific to the saccharimeter used), conforming to ICUMSA class A specifications.
Analytical balance with an accuracy of ± 0.001 g.
Water bath, thermostatically controlled to 20.0 ± 0.1°C.

Reagents: Distilled or deionized water.

Procedure: The procedure is applicable to white sugars with low color and turbidity. Sugars which are too colored or too turbid for direct polarization are polarized after purification with basic lead acetate (according to raw sugar polarization procedures).
The normal weight (26 ± 0.002 g) of the sugar sample is transferred to a 100-ml flask by washing with distilled or deionized water (about 80 ml). The sugar is dissolved by agitation and heating, and water is added to just below the calibration mark. The temperature is adjusted to 20 ± 0.1°C by means of the water bath. The inside wall of the neck of the flask is dried with filter paper, and the solution volume is adjusted to exactly 100 ml with 20 ± 0.1°C water.
The flask is then sealed with a clean, dry stopper, and its contents are mixed thoroughly by hand-shaking.
The polarimeter tube is rinsed twice with approximately ⅔ of the sugar solution to be tested. The tube is then filled with the test sugar solution at 20 ± 0.1°C in such a way that no bubbles are entrapped. The tube is placed in the saccharimeter and polarized at 20 ± 0.1°C. Five measurements are taken to 0.05°S or better. The average value is expressed to the nearest 0.01°S. In the

same way, the quartz control plate reading is determined to 0.01°S.

For saccharimeters without quartz-wedge compensation, no correction is necessary if the above procedure is carried out as described. With quartz-wedge saccharimeters, it is not normally convenient to adjust the temperature of the quartz-wedge compensator to 20 ± 0.1°C, and consequently, the following equation is used to apply the necessary correction:

$$P_{20} = P_t\,[1 + 0.00014\,(t_q - 20)]$$

where: P_{20} = polarization at 20 ± 0.1°C
P_t = observed polarization
t_q = temperature in ° C of the quartz-wedge compensator.

Flask Correction: If a flask correction is required, then the polarization is corrected by adding the flask correction to the observed polarization.

Flask Correction = Actual volume of flask – 100.00.

Calculation: The result is expressed as polarization in °S to an accuracy of 0.01°S.

Method Number: USNC MOA 0010

Analysis: Specks

Purpose: The quantity of foreign material or specks found in white sugar is determined by visual inspection.

Apparatus: Board (2 ft x 2 ft) framed 2 inches high and painted high-gloss white.
Balance, 1000 g capacity minimum.

Procedure: From a well-mixed representative sample, weigh out approximately 1000 g and place on the white board. Spread the sugar over the entire surface of the board. Count any visible specks.

Calculation: $\text{Visible specks/100 g} = \frac{\text{Visible specks counted}}{10}$

18

Sugar Industry Terminology

R. L. Knecht*

The listing below is intended to briefly describe the terminology encountered in discussing sugar and its applications. It is intended only as a general reference and not as a full technical definition of the terms involved. Some words have different usages in different applications, and where such dual meanings are common, they are indicated.

Activated Carbon—An absorbent, formed by the retorting of carbon materials at very high temperatures to expose large internal surface areas, used to decolorize sugar liquors.

Affination—The process of "washing" the surrounding syrup from crystallized sugar. Affination is a centrifugal separation of liquid syrup and solid crystals.

Agglomerated Sugar—Attached groupings of sucrose microparticles to form free-flowing granules.

Airslide—A specific type of bulk carrier equipped with a porous, inclined bottom through which air fluidizes the discharge of the contained commodity.

AOAC—Abbreviation for Association of Official Analytical Chemists.

Apparent Purity—The pol of a solution divided by its Brix.

Ash—The residue remaining after total combustion of a solution or mixture; used as a measure of the inorganic components of the starting material. Generally used to refer to total inorganic constituents.

*R. L. Knecht is Senior Vice President, Refinery Operations, C & H Sugar Company, Crockett, CA.

Attenuation Index—The color measure for the absorption of light at a specific wavelength through a solution.

Attrition—The breakage of sugar crystals during transport that results in the generation of dust (fines).

Bacteria—Any of a class of microscopic plants having various shapes, sizes, motilities, and environmental specializations.

Bagasse—The crushed plant fiber remaining after the extraction of the sugar-containing juice from sugarcane. A source of cellulose and fuel.

Bakers Sugar—A refined specialty product that has an average crystal size smaller than that of normal table sugar.

Bale—A variety of external shipping containers that cover 2-, 5-, or 10-lb bags to form a shipping unit of 60 lb net.

Baumé—A specific gravity measurement scale based on a hydrometer measurement of solids in solution.

Beet Sugar—Sucrose processed from the sugar beet plant.

Blackstrap—A type of molasses which is generally used as animal feed or biological (fermentation) feed stock. The by-product of sugar extraction from sugar-containing liquors.

Bone Char—A decolorizing and de-ashing filtration material made from dried animal bones that have been crushed and retorted to activate their alkaline calcium phosphate crystalline structure and carbon. The bone char absorbent can be reactivated and reused after retorting at 550°C.

Bottlers Sugar—A refined specialty product formulated to meet the requirements of beverage bottlers. A low-color product that meets exacting ash and microbiological specifications.

Brix—The percent by weight of sucrose in a water solution. °Brix can be determined by measuring the specific gravity or refractive index of a solution and comparing either of these measurements to the appropriate tabulated data of pure sucrose and water solutions.

Brown Sugar—A finished sugar product consisting of sugar crystals and darker nonsucrose materials. Soft brown sugars are brown sugars which are crystallized directly by specialized crystallization processes. Brown sugars can also be made by blending white crystallized sugar and dark syrups.

Bulk Density—The weight per unit volume of a large mass of material. Bulk density values are purely comparative for the conditions that apply. Bulk density is reported as lb/ft^3 for solids and lb/gal for liquids.

Bulk Sugar—A refined product sold in bulk, not packaged, form. Usually these products are loaded into tanker railcars, trucks, or into hoppers or containers for shipment.

Bundle—One of a variety of external shipping wraps for 2-, 5-, or 10-lb sugar bags to form a shipping unit of 60 lb net.

Caking—The hardening of sugar due to loss of moisture.

Cane Sugar—Sucrose processed from the sugarcane plant.

Canners Sugar—A refined, granulated specialty product that meets exacting standards for microbiological quality.

Caramel—The product of heating sucrose to decomposition. Also a type of confectionery product.

Carbohydrate—A family of organic compounds composed of carbon, hydrogen, and oxygen atoms in the ratio of 1:2:1, respectively.

Centrifugal/Centrifugation—The equipment/process used to separate liquids and solids by spinning the mixtures at high speeds. The liquid is separated from the solid by filtration through a screen. This separation is enhanced by centrifugal force.

Clarifier—A piece of equipment whose major purpose is to remove suspended solids and/or colloidal materials from a liquid. As applied to sugar, these are normally either flotation or sedimentation devices.

Clerget—An analytical procedure to determine true sucrose content in the presence of invert sugar (glucose and fructose).

Color—The property of a material in which specific wavelengths of visible light are absorbed.

Conductivity/Conductivity Ash—The measurement of the electrical conductance of a solution. Conductance is correlated with the amount of inorganic salts present.

Conglomerate—Multiple crystals cemented together.

Confectioners Sugar—Refined sugar products whose granulations range from coarse to powdered. Granulated sugar products processed to meet specific applications, and include coarse, sanding, extra-fine, fruit, bakers special, and powdered sugars.

Corn Sugar—The sugar products derived from the acid or enzyme hydrolysis of corn starch. These starch conversion products are normally liquids of varying glucose (dextrose) content. The initial product can be processed further to yield a higher level of fructose (levulose).

Crystal/Crystallization—The ordered arrangement of molecules within a solid form. The process of forming crystals.

CV—Abbreviation for coefficient of variance that is used in conjunction with MA (mean aperture) to describe the crystal size distribution of granulated and powdered sugar products.

D.E.—Abbreviation for dextrose equivalent; D.E. is a measure of the total reducing sugars, expressed as dextrose, of corn syrups.

Decolorization—The process of removing colored impurities from sugar solutions by absorption with bone char, activated carbon, and/or ion-exchange resin.

Demurrara—A dark brown specialty sugar product.

Density—The weight per unit volume of a discrete particle of matter. For example, the density of a single sugar crystal is reported as g/ml.

Dextrose—A monosaccharide, also called glucose or blood sugar, having the chemical formula $C_6H_{12}O_6$. Dextrose is one of the two component monosaccharides of sucrose.

DNS—Abbreviation for dry nonsucrose; total amount of dry nonsucrose materials in a mixture or solution. Generally equivalent to 100 minus true purity.

Evaporation—The removal of water as its vapor. In sugar processing, this is normally done under a controlled vacuum to boil away water at lower temperatures and higher energy efficiencies.

Factory—A sugar beet processing facility in which refined sugar is recovered from harvested sugar beets; or, a facility in which harvested sugarcane is processed into raw sugar.

Fermentation—A biological process that involves the microbial conversion of sugar by yeast or other organisms. In the case of yeast, sugar is converted to alcohol and carbon dioxide.

Filtration—The process of separating a solid from a liquid by applying a force to move the liquid through a barrier while retaining the solid.

Fine Sugar (Fine Granulated)—A refined sugar product made up of sugar crystals whose average size and distribution fall into the range of normal table sugar. Also called extra-fine granulated sugar by some producers.

Floc—A precipitate that remains suspended in a solution.

Fondant—A creamy preparation of fine sugar crystals; a nearly homogeneous suspension used as a basis for icings and candies. A type of powdered sugar.

Fructose—A monosaccharide, also called levulose or fruit sugar, having the chemical formula of $C_6H_{12}O_6$. Fructose is one of the two component monosaccharides of sucrose.

Glucose—A monosaccharide, also called dextrose or blood sugar, having the chemical formula $C_6H_{12}O_6$. Glucose is one of the two component monosaccharides of sucrose.

Grain Size—The size of a crystal as measured by the smallest sieve opening through which it can pass.

Granulated Sugar—Refined white, crystalline sucrose.

HFCS—Abbreviation for high fructose corn syrup, a sweetener prepared by hydrolyzing corn starch into glucose followed by enzymatic conversion of a portion of the glucose into fructose.

Hydrolysis—The chemical reaction that involves the addition of a water molecule to the reactant(s). The conversion of sucrose into invert sugar is hydrolysis, as a molecule of water is added to the sucrose molecule.

Hydrometer—A device for measuring specific gravity. The hydrometer is immersed in a solution and the relative displacement within the solution is compared to that within pure water.

Icing Sugar—A fine powdered sugar product.

ICUMSA—Abbreviation for the International Commission for Uniform Methods of Sugar Analysis.

Invert Sugar/Inversion Reaction—The product of the hydrolysis of sucrose to yield an equal mixture of glucose (dextrose) and fructose (levulose).

Invertase—The enzyme that catalyzes the hydrolysis of sucrose to glucose and fructose.

Ion-exchange—A process that uses specially fabricated porous beads which are chemically modified to exchange one ion for another as a solution is passed through them.

Juice—The fluid expressed from the sugar beet root or sugarcane stalk, containing the dissolved sucrose formed by the plant.

Kieselguhr—A filtration aid. Also called diatomaceous earth.

Levulose—A monosaccharide, also called fructose or fruit sugar, having the chemical formula $C_6H_{12}O_6$. Levulose is one of the two component monosaccharides of sucrose.

Liquid Sugars—Finished sugar products sold in the liquid state as a concentrated solution of the particular sugar product and water.

Liquor—A high-Brix, sugar-containing solution recovered at various stages of the refining process.

Magma—A mixture of sugar crystals and liquid that forms a thick slurry.

Massecuite—Magma produced in a vacuum pan. The suspension of sugar crystals in their mother liquor produced during the first stages of crystallization.

Mean Aperture—The average opening through which a collection of particles will pass. Used to describe the average size of sugar crystals.

Melt—Used in refining terminology to mean "dissolve." A refinery's melt is a measure of the total intake of raw sugar.

Mill—The term applied to a cane sugar factory where sugarcane stalks are converted into raw sugar, or to the equipment that mechanically crushes and pulverizes the sugarcane stalks prior to extraction of the sugar-containing juices.

Mingle—The mixing of sugar crystals and a liquid in such proportions that the crystals are not dissolved to form a magma or thick slurry.

Molasses—The by-product of sugar extraction. The exhausted mother liquid of the crystallization process. A thick syrup whose purity has been reduced to the point that further crystallization of sugar is uneconomical.

Nonnutritive Sweetener—A sweetening agent that does not possess significant nutritive properties in proportion to its sweetness. Artificial sweeteners are generally nonnutritive at their rates of usage.

Nutritive Sweetener—A sweetener that contributes caloric value at its usage rate. All natural carbohydrates are nutritive.

Number 11 Contract—The trading unit for the world sugar futures on the New York Coffee, Sugar & Cocoa Exchange.

Number 14 Contract—The trading unit for domestic sugar futures on the New York Coffee, Sugar & Cocoa Exchange.

Oligosaccharide—A sugar polymer that results when 3 to 10 monosaccharide molecules are linked chemically.

Organic Compound—Natural or manmade material that is based on the element of carbon. Organic compounds are the result of reactions that cause carbon to be linked to other elements.

Pallet—Rigid framework on which packaged sugar is unitized for storage and shipping. In most sugar applications, a standard Grocery Manufacturer's Association (GMA) 40- by 48-inch, 4-way entry pallet is used.

Pan/Vacuum Pan—The piece of equipment used in sugar processing that allows the controlled boiling of solutions under a vacuum.

pH—The measurement of hydrogen ion activity of a solution. A pH above 7 refers to an alkaline solution; a pH below 7 refers to an acidic solution.

Pneumatic Trailer—A type of bulk carrier that allows unloading of the product by air pressure.

Pol/Polarization—Used interchangeably as a measure of the optical rotation of a solution under standardized (ICUMSA) conditions, expressed as a percentage of pure sucrose in a standard solution.

Polarimetry/Polarimeter—The amount of rotation experienced by a beam of polarized light as it passes through a solution. When used to measure the sucrose content of a sugar solution, it is termed saccharimetry. The amount of sucrose determined by this method is called the pol of the solution. The instrument used to measure this rotation is called a polarimeter (also called polariscope).

Polysaccharide—A sugar polymer that results when ten or more separate monosaccharide molecules are linked chemically.

Powdered Sugar—A sugar product produced by grinding a mixture of granulated sugar and corn starch.

Pressed Sugar—A sugar product formed by molding damp granulated sugar into shapes such as cubes, and subsequent drying of the product.

Purity—The amount of sucrose in a material divided by the total amount of dry substance present. Usually expressed as a percentage.

RDS—Abbreviation for refractometric dry substance, expressed as the °Brix of a solution as measured by a refractometer.

Raw Sugar—The product of cane sugar factories or mills. An intermediate crystalline product resulting from the evaporation of water from sugarcane stalk juice.

Raw Value—The expression of the amount of sugar converted to the equivalent amount of 96-Pol raw sugar.

Reducing Sugar—Sugars that act as reducing agents. These highly reactive sugars include glucose and fructose, and this term is often used to represent their presence in sugar analysis.

Refractometer/Refractometric Solids—An instrument for measuring the refractive index of a solution. A measurement of the total solids in solution, expressed as °Brix when compared to a pure sucrose-water solution.

Refractive Index—The relative bending, refraction, of a beam of light of defined wavelength as it passes through a solution as compared to passing through pure water.

Sugar Refinery—A processing facility in which nonsugar impurities are removed from raw sugar to produce a variety of sugar products for human consumption.

Reflectance Color—A measurement of the color of a material by the relative amount of light reflected off its surface.

Saccharimeter—A polarimeter used specifically to measure the optical rotation of sucrose solutions. A saccharimeter is calibrated to read optical rotation in °S or °Z.

Sanding Sugar—A refined specialty product that has an average crystal size much larger than that of normal table sugar. Normally used to decorate cookies, candies, etc.

Saturation—The level at which a solution contains all the material it is capable of dissolving under current conditions.

Sediment—The solid material which settles out from a fluid. This term is also used to indicate the amount of solid, undissolved impurities present in a solution.

Soft Sugar—A brown sugar product prepared by a special crystallization process. Soft sugars are sugar crystals containing a film of remaining mother liquor and are noted for their unique flavor, color, and functional properties.

Solubility—The relative ability of a material (solute) to be dissolved by another (solvent).

Specialty Sugars—Those sugar products with unique specifications tailored for specific applications. These products normally include all products other than the usual granulated sugars.

Specific Gravity—The weight of a volume of material divided by the weight of that same volume of water.

Specific Heat—The amount of energy required to raise the temperature of one gram of a material one degree Celsius compared to the energy required to raise the temperature of one gram of water one degree Celsius. Individual energy requirements are reported in calories; the dimensions of specific heat are unity.

Starch—A polymer consisting of long chains of linked glucose molecules. Starch contains both branched (amylopectin) and unbranched (amylose) components.

Strike—The contents of a vacuum pan. The act of starting the crystallization process in a vacuum pan.

Sugar/Sucrose—The natural product of photosynthesis in green plants. A disaccharide molecule with the chemical formula of $C_{12}H_{22}O_{11}$, consisting of one molecule each of glucose and fructose.

Syrup—A viscous, concentrated sugar solution resulting from the evaporation of water. Also the remaining mother liquor after crystallization of sugar from a solution.

Tablets—Products that result when damp sugar is pressed and subsequently dried (see Pressed Sugars).

Total Sugars—A measurement of the total amount of saccharides present, which includes sucrose, glucose, and fructose. Most often used in conjunction with molasses.

Turbidity—The cloudiness of a fluid as measured by the amount of light that is scattered when light is passed through the fluid.

Turbinado—A semirefined light brown crystalline sugar.

USP—Abbreviation for United States Pharmacopoeia and a listing of quality standards for various products including sugar.

Viscosity—The resistance of a liquid material to flow.

Yeast—A microorganism that ferments sugar; the most common products are carbon dioxide and ethyl alcohol.

19

The Sugar Industry

Theodore Cayle, Ph.D. *

Following is a listing of refined sugar producers, both beet and cane, in the United States and Canada; cane raw sugar producers in the United States; sugar industry affiliates in the United States and Canada; and foreign sugar affiliates.

A. Refined Sugar Producers

1. USA

Amalgamated Sugar Company
First Security Bank Building
P. O. Box 1520
Ogden, UT 84402
Ph: 801/399-3431
Fax: 801/399-3431, ext. 301
Tlx: 910/971-5914

American Crystal Sugar Company
101 North Third Street
Moorhead, MN 56560
Ph: 218/236-4400
Fax: 218/236-4494

Amstar Sugar Corporation
1251 Avenue of the Americas
New York, NY 10020
Ph: 212/489-9000
Fax: 212/302-2794

California and Hawaiian Sugar Company
1390 Willow Pass Road
Concord, CA 94520
Ph: 415/356-6000
Fax: 415/356-6037

Colonial Sugars, Inc. (Division of Savannah Foods & Industries, Inc.)
P. O. Box 1646
First National Bank Building
Mobile, AL 36633
Ph: 205/433-0454
Fax: 205/431-1208

*The late Theodore Cayle, Ph.D., was Vice President of Scientific Affairs, The Sugar Association, Inc., Washington, DC.

Delta Sugar Corporation
P. O. Box 58
Clarksburg, CA 95612
Ph: 916/744-1711
Fax: 916/744-1516

Great Lakes Sugar Company (Division of Michigan Sugar Company)
4800 Fashion Square Boulevard
Plaza North, Suite 300
P. O. Box 1348
Saginaw, MI 48605
Ph: 517/799-7300
Fax: 517/799-7310

Holly Sugar Corporation (Division of Imperial Holly Corporation)
P. O. Box 1052
Colorado Springs, CO 80901
Ph: 719/471-0123
Fax: 719/630-3252

Imperial Holly Corporation
P. O. Box 9
Sugar Land, TX 77487
Ph: 713/491-9181
Fax: 713/491-9198

Michigan Sugar Company (Division of Savannah Foods & Industries, Inc.)
4800 Fashion Square Boulevard
P. O. Box 1348
Saginaw, MI 48605
Ph: 517/799-7300
Fax: 517/799-7310

Minn-Dak Farmers Cooperative
Route 1 Box 10
Wahpeton, ND 58075
Ph: 701/642-8411
Fax: 701/642-6814

Monitor Sugar Company
2600 South Euclid Avenue
Bay City, MI 48706
Ph: 517/686-0161
Fax: 517/686-7410

Refined Sugars, Inc.
1 Federal Street
Yonkers, NY 10702
Ph: 914/963-2400
Fax: 914/963-1030

Savannah Foods & Industries, Inc.
P. O. Box 339
Savannah, GA 31402
Ph: 912/234-1261
Fax: 912/238-0252

Southern Minnesota Sugar Cooperative
East Highway 212
P. O. Box 500
Renville, MN 56284
Ph: 612/329-8305
Fax: 612/329-3252

Spreckels Sugar Company, Inc.
4256 Hacienda Drive
P. O. Box 9025
Pleasanton, CA 94566-4065
Ph: 415/463-3400
Fax: 415/463-3714

Supreme Sugar Company Inc. (Division of ADM Company)
111 Veterans Boulevard
Metairie, LA 70005
Ph: 504/831-0901
Fax: 504/831-0909

The Western Sugar Company
1700 Broadway
Denver, CO 80290
Ph: 303/830-3939
Fax: 303/830-3940

2. Canada

British Columbia Sugar Refining Company, Ltd.
123 Rogers Street
P. O. Box 2150
Vancouver, British Columbia
V6B 3V2
Ph: 604/253-1131
Fax: 604/253-2517

Lantic Sugar Limited
1 Westmount Square
Montreal, Quebec H3Z 2P9
Ph: 514/939-3939

Redpath Sugars
P.O. Box 66
Royal Bank Plaza
Toronto, Ontario M5J 2J2
Ph: 416/360-6266
Fax: 416/360-7379

B. Cane Raw Sugar Producers—USA

1. Florida

Atlantic Sugar Association
P. O. Box 1570
Belle Glade, FL 33430
Ph: 407/996-6541
Fax: 407/996-8021

Okeelanta Corporation
P. O. Box 1059
Palm Beach, FL 33480
Ph: 407/655-6305
Fax: 407/659-3206

Osceola Farms Company
P. O. Box 1059
Palm Beach, FL 33480
Ph: 407/655-6305
Fax: 407/659-3206

Sugar Cane Growers Cooperative of Florida
Airport Road
P. O. Box 666
Bell Glade, FL 33430
Ph: 407/996-5556
Fax: 407/996-4747

United States Sugar Corporation
111 Ponce De Leon Avenue
P. O. Box 1207
Clewiston, FL 33440
Ph: 813/983-8121
Fax: 813/983-4804

2. Hawaii

Gay & Robinson, Inc.
Makaweli, HI 96769
Ph: 808/338-8233

Hamakua Sugar Company, Inc.
P. O. Box 250
Paauilo, HI 96776
Ph: 808/537-2505
Fax: 808/776-1250

Hawaiian Commercial & Sugar Company
P. O. Box 266
Puunene, HI 96784
Ph: 808/877-0081

Hilo Coast Processing Company
P. O. Box 18
Pepeekeo, HI 96783
Ph: 808/963-5516

Ka'U Agribusiness Company, Inc.
P. O. Box 130
Pahala, HI 96777
Ph: 808/928-8311

Kekaha Sugar Company, Ltd.
P. O. Box 549
Kekaha, HI 96752
Ph: 808/337-1472

The Lihue Plantation Company, Ltd.
P. O. Box 751
Lihue, HI 96766
Ph: 808/245-2112

Mauna Kea Agribusiness Company, Inc.
P. O. Box 68
Papaikou, HI 96781
Ph: 808/964-1025

McBryde Sugar Company, Ltd.
P. O. Box 8
Eleele, HI 96705
Ph: 808/335-5333

Oahu Sugar Company, Ltd.
P. O. Box 0
Waipahu, HI 96797
Ph: 808/677-3577

Olokele Sugar Company, Ltd.
P. O. Box 156
Kaumakani, HI 96747
Ph: 808/335-5337

Pioneer Mill Company, Ltd.
P. O. Box 7272
Lahaina, HI 96761
Ph: 808/661-0592

Waialua Sugar Company, Inc.
P. O. Box 665
Waialua, HI 96791
Ph: 808/637-6284

Wailuku Agribusiness Company, Inc.
P. O. Box 520
Wailuku, HI 96793
Ph: 808/244-7079

3. Louisiana

Alma Plantation, Ltd.
Lakeland, LA 70752
Ph: 504/627-6666

Breaux Bridge Sugar Corporation
P. O. Box 236
Breaux Bridge, LA 70517
Ph: 318/332-2231

Caire & Graugnard
P. O. Box 7
Edgard, LA 70049
Ph: 504/446-1084

Cajun Sugar Cooperative, Inc.
P. O. Box 1179
New Iberia, LA 70560
Ph: 318/365-3401

Caldwell Sugars Cooperative, Inc.
P. O. Box 226
Thibodaux, LA 70301
Ph: 504/447-4023

Cora-Texas Manufacturing Company, Inc.
P. O. Box 280
White Castle, LA 70788
Ph: 504/545-3679

Dugas & LeBlanc, Ltd.
P. O. Box 10
Paincourtville, LA 70391
Ph: 504/473-4620

Evan Hall Sugar Cooperative, Inc.
P. O. Box 431
Donaldsville, LA 70345
Ph: 504/473-8241

Glenwood Cooperative, Inc.
P. O. Box 545
Napoleonville, LA 70390
Ph: 504/369-2943

Iberia Sugar Cooperative, Inc.
P. O. Box 11108
New Iberia, LA 70562
Ph: 318/364-1913
Fax: 318/365-0030

Jeanerette Sugar Company, Inc.
2304 West Main Street
Jeanerette, LA 70544
Ph: 318/276-4238

Lafourche Sugar Corporation
P. O. Box 551
Thibodaux, LA 70301
Ph: 504/447-3210

Harry L. Laws & Company, Inc.
P. O. Box 158
Brusly, LA 70719
Ph: 504/749-2861

M. A. Patout & Son, Ltd.
Route 1, Box 288
Jeanerette, LA 70544
Ph: 318/276-4592

Savoie Industries, Inc.
P. O. Box 68
Belle Rose, LA 70341
Ph: 504/473-9293

South Coast Sugars, Inc.
P. O. Box 159
Raceland, LA 70394
Ph: 504/523-5518

St. James Sugar Cooperative, Inc.
St. James, LA 70086
Ph: 504/265-4056

St. Martin Sugar Cooperative, Inc.
Route 1, Box 1038
St. Martinville, LA 70582
Ph: 318/394-3255

St. Mary Sugar Cooperative, Inc.
P. O. Box 269
Jeanerette, LA 70544
Ph: 318/276-6761

Sterling Sugars, Inc.
P. O. Box 572
Franklin, LA 70538
Ph: 318/828-0620

4. Texas

Rio Grande Valley Sugar Growers, Inc.
P. O. Drawer A
Santa Rosa, TX 78593
Ph: 512/636-1411
Fax: 512/636-1046

C. Sugar Industry Affiliates

1. USA

American Society of Sugar Beet Technologists
2301 Research Boulevard
Fort Collins, CO 80526
Ph: 303/482-8250

American Society of Sugar Cane Technologists
Knapp Hall
Louisiana State University
Baton Rouge, LA 70803
Ph: 504/388-2467

American Sugarbeet Growers Association
1156 15th Street, NW
Suite 1020
Washington, DC 20005
Ph: 202/833-2398
Fax: 202/785-7312

American Sugar Cane League of the USA, Inc.
206 E. Bayou Rd.
P. O. Box 938
Thibodaux, LA 70302-0938
Ph: 504/448-3707
Fax: 504/448-3722

Beet Sugar Development Foundation
P. O. Box 1546
Fort Collins, CO 80522
Ph: 303/482-8250

Florida Sugar Cane League, Inc.
P. O. Box 1208
Clewiston, FL 33440
Ph: 813/983-9151
Fax: 813/983-2792

Hawaiian Sugar Planters' Association
99-193 Aiea Heights Drive
P. O. Box 1057
Aiea, HI 96701
Ph: 808/487-5561
Fax: 808/486-5020

Sugar Processing Research, Inc.
1100 Robert E. Lee Boulevard
New Orleans, LA 70124
Ph: 504/286-4542
Fax: 504/282-5387

The Sugar Association, Inc.
1101 15th Street, NW
Suite 600
Washington, DC 20005
Ph: 202/785-1122
Fax: 202/785-5019

United States Cane Sugar Refiners Association
1001 Connecticut Avenue, NW
Suite 735
Washington, DC 20036
Ph: 202/331-1458
Fax: 202/785-5110

United States Beet Sugar Association
1156 15th Street, NW
Suite 1019
Washington, DC 20005
Ph: 202/296-4820
Fax: 202/331-2065

United States National Committee for Sugar Analysis
1100 Robert E. Lee Boulevard
New Orleans, LA 70719
Ph: 504/286-4542
Fax 504/282-5387

2. Canada

Canadian Sugar Institute
7 King Street East, Suite 1902
Toronto, Ontario M5C 1A2
Ph: 416/368-8091
Fax: 416/368-6426

Sugar Industry Technologists
P. O. Box 632
Ste. Thérèse de Blainville
Quebec, J7E 4KE Canada
Ph: 514/621-3524

3. Foreign

CEFS
European Committee of Sugar Manufacturers
182 Avenue de Tervueren
B1150 Brussels, Belgium
Ph: 011-32-2-762-0760
Fax: 011-32-2-771-0026

CIBE
International Confederation of European Beet Growers
29, rue du Général Foy
F-75008 Paris
France
Ph: 42 94 41 00

CITS
International Commission of Sugar Technology
1 Aandorenstraat
B-3300 Tienen
Belgium
Ph: 02/81 30 11

Commission of the European Communities, Sugar Division
Rue de la Loi 200
B-1029 Bruxelles
Belgium
Ph: 02/235 11 11

GEPLACEA
Group of Latin American and Caribbean Sugar Exporting Countries
Av. Ejecito Nacional 373–lo. piso
Mexico DF 11520
Ph: 05/250-7566

ICUMSA
International Commission for Uniform Methods of Sugar Analysis
Box 5611
Mail Centre
Mackay 4741, Australia
Ph: 079/521-511
Fax: 079/521-734

IIRB
International Institute for Sugar Beet Research
Rue Montoyer, 47
B-1040 Bruxelles
Belgium
Ph: 02/512 65 06

International Association for Sugar Statistics
P. O. Box 1220
Am Muhlengraben 22
D-2418 Ratzeburg
West Germany
Ph: 04541/8 34 03

ISO
International Sugar Organization
28 Haymarket
London SW1Y 4SP
United Kingdom
Ph: 01/930 3666

World Association of Beet and Cane Growers
21 rue Chaptal
75009 Paris
France

WSRO
World Sugar Research Organisation
University of Reading
Innovation Centre
Philip Lyle Bldg.
P. O. Box 68
Reading RG6 2BX
United Kingdom
Ph: 01-44-7-34861361
Fax: 01-44-7-34312198

Index

Acidity. *See also* pH
  effect on sucrose, 57, 109, 166
  in gelling, 216, 218
  in sucrose inversion, 200
Advertising, in cereals industry, 184
Aerated icings, 146-148
  formulas, 148
Aerating agents, 112-113, 140
Aeration, mechanical, 122-123
After-dinner mints, formula, 116-117
Agglomeration processes, 59, 169-170
Albumin
  as aerating agent, 112-113
  as purifying agent, 22
Alcoholic beverages, sugar in, 210-211
Alkalinity. *See also* pH
  effect on sucrose, 57, 109, 166
Almond crunch, microwave recipe, 261-262
American Chemical Society, 278
American Council on Science and Health, evaluation of sugar and health, 100
American Dietetic Association, 99
American Society for Clinical Nutrition, Inc. Task Force Report, 99
Analytical methods, for white sugar, 288-302
Angel food cakes, 142
  formula, 143
Angle of repose, of sucrose solution, 54
Anticaking agents, 177
Appearance
  icings and, 145
  of pie fillings, 175
  of processed foods, 165-167
Apple jelly formula, 223
Aroma enhancement, by sugar, 140, 171, 209
Artificial colors, in preserves, 216
Artificial sweeteners, comparison with natural sweeteners, 65, 74
Ascorbic acid, as antioxidant, 171
Ash
  analytical method for, 293
  in brown sugars, 40
Atherosclerosis studies, 89
Attenuation factor
  dielectric, 241
  equation for, 269

Bacon, sugar-curing of, 180
Bagasse, 17
Bakers special sugar, 37
Bakers yeast, 134-135
Bakery foods, sugar role in, 130-151
Baking industry, 132, 150
Baking powder, 143
Baumé scale, 49
Beaters, for candymaking, 111
Beet sugar. *See also* Sugar
  cultivation, 26-29
  factories, 34
  geography, 27
  harvest operations, 28-29
  history, 5-7
  industry, 26-35
  insect pests, 28

photosynthesis in, 83-84
processing, 30-33
processing flow diagram, 31
storage, 29-30
yield, 27-29, 34
Behavior, neurotransmitter levels and, 95-96
Beverages
carbonated, 198-209
fruit-flavored powders for, 172-173
grapefruit pureé, 174
sugar role in, 172-174, 198-211
Binding agent, sugar as, 188, 191, 285
Biological control, for cane sugar, 17
Biosynthetic production, of sucrose, 84-85
Biotechnological uses, of sugar, 58, 281-283
Blood lipids, sucrose intake and, 94
Boiled icings, 145
Boiling point, of sucrose solution, 53, 104-106
Bottlers standards, 208
Bread flavor, sugar and, 136
Breadmaking, 132-139
comparison of procedures, 133
Breakfast cereals. *See* Cereals
Breeding programs, 14-16
British Food Manufacturing Industries Research Association, 277
British Nutrition Foundation Task Force Report, 99
Brix scale, 49, 199-200
Brown sugars, 40
analysis, 43
in baking, 131
flavor, 69
properties, 59
storage and handling, 62
Browning, 166-167
in baking, 136, 143-144
in cereals, 188, 192
inhibition of, 171, 176, 210-211
in microwave cooking, 234
Bulk density, in microwave cooking, 249, 251
Bulking agent, sugar as, 165, 169-170
Bulky flavors, of ice cream, 155
Butter crunch, formula and procedure, 118
Buttercream icings, formula and procedure, 148

Cakes
formulas, 140, 143
shortening in, 140-142
sugar in, 137, 139-143
Candy. *See* Candymaking. *See also* specific candies
Candymaking, sugar role in, 103-128
Cane raw sugar producers, 314-316
Cane sugar. *See also* Sugar
borers, 16
cultivation, 11-17
diseases, 16
field testing, 14-16
genetic differences among plants, 15-16
harvest operations, 13-14
history, 1-5, 212-213
insect pests, 16-17
photosynthesis in, 85
pineapple disease, 16
processing, 11-25
ratoon stunting disease, 16
refining, 23-25
yield, 7-9
Carageenan, as thickening agent, 158, 175
Caramel flavor, 160
Caramelization. *See* Browning
Caramels, 124-125
formula and procedure, 126-127
Carbohydrates
chemistry, 47
sensitivity to, 92

Carbon dioxide, in fermentation, 134
Carbonated beverages, 198-211
Carbonation. *See* Purification
Carcinogenicity studies, 89
Cardiovascular disease, sugar consumption and, 95
Carrot cake, microwave recipe, 258-259
Cast creams, 114
  formula, 115
Central nervous system, sugar and, 85
Centrifugation, 20-22, 31
Cereal industry, 182-184
Cereals
  coatings for, 185, 187-188, 190
  comparison of ingredients in, 193-194
  manufacturing processes for, 184-187
  sugar in, 182-197
Charcoal filtration, 22
Chemical reactions, of sucrose, 278-281
Chemically leavened bakery foods, 139-145
Chewiness
  of candies, 122-126
  of ice cream, 156
Chlorinated cake flours, 141
Chocolate cake, microwave recipe, 258
Chocolate frosting, microwave recipe, 258
Chocolate fudge, microwave recipe, 261
Chocolate mousse brownies, microwave recipe, 259
Chocolate-flavored milk products, 157-158, 160, 169-170
  formula, 158, 170
Chromatography, in flavor analysis, 69
Citrus pureé base, formula, 174
Clarification, 17, 19, 32
Coated brown sugar, 40
Coatings. *See* Surface coatings. *See also* Frostings, Icings
Coffee fudge, microwave recipe, 261
Cold process, for icings, 145
Color
  additives, 177
  analytical method for, 289
  in bakery foods, 131
  of brown sugars, 40
  of molasses, 41
  in preserves and jellies, 219-220
  purity and, 36, 39, 62, 210-211
  sugar as enhancer of, 132, 140, 165-167, 171
Combustibility, of sugar, 61
Commercial baking methods, 143
Complex permittivities, 239-240
  equation for, 269
  of solutions, 237
  of sugars, 249
  of wheat flour, 251
Condensed milk. *See also* Sweetened condensed milk
  in ice cream, 153
Conditioning, of sugar products, 38
Conductivity. *See* Electrical properties
Confectionery. *See* Candymaking
Conveyers, of brown sugar, 63
Cookies
  fillings, 149-150
  formulas, 144
  procedure, 150
  sugar in, 139, 143-145
  surface cracking, 146
Cooling. *See also* Temperature
  in candymaking, 111
Cooperatives, 5
Copper, effect on flavor, 122
Cordials, sugar in, 210-211
Corn flakes, formula, 189
Corn starch
  in confectioners sugar, 39
Corn sweeteners
  in beverages, 198
  in candymaking, 110, 113, 119-121

in ice cream, 152-153, 155
in icings, 148-149
Cosmetics, sucrose in, 284
Cream fillings, formulas and procedure, 149-150
Cream of tartar
in cake baking, 143
in candymaking, 116-117
Creaming, 106-107, 111, 141, 144
Creams
in ice cream, 153
production of, 112-116
Crumb texture, sugar and, 137-138
Crystallization, 19-20, 30, 33
in candymaking, 106
of cereal frostings, 192
control in icings, 146, 149
early methods for, 6
of granulated sugar, 38
retardation of, 40
Cultivation methods, 11-16, 26, 28-29
Cutting machine cookies, 144

Dairy products
chocolate-flavored, 157-158
frozen novelties as, 156
ice cream and ice milk, 153-155
ices, 157
sugar role in, 152-164
Danish sweet doughs. *See* Sweet doughs
Density of foods
complex permittivity and, 251
dielectric properties and, 249
Dental caries
fluoride reduction of, 91
sugar consumption and, 89, 90-91, 98-99
Surgeon General's Report and, 98
Deposit cookies, 144
Dextran, 282-283
Diabetes mellitus, glucose metabolism and, 88, 93-94
Diatraea sacharalis, 16
Dielectric properties
of alcohols, 246-248
of foods, 236, 238-241, 243-249, 267
of salts and ions, 245
of sugars, 228-229, 246-248
of water, 243-245, 272
Dietary Goals for the United States, 98
Diffusion, in beet sugar processing, 28, 30
Dilutant, sugar as, 165
Directory, 312-318
Diseases
beet sugar, 28
cane sugar, 16
Dispersion,
solutions and, 246
sugar role in, 165, 170
Drinks. *See* Beverages
Dry mixes, 37
for beverages, 170, 172-173, 177, 210
Drying oil esters, sucrose in, 280

Egg nog products, 159-160
formula, 160
Egg whip. *See* Frappé
Electrical properties
conductivity equation and, 271
of foods, 252-253
of sucrose solution, 55
Electromagnetics, microwave, 231-236
Emulsifiers, in ice cream, 153
Energy
equation, 269
microwave, 230-235
Energy production, sugars and, 87
Enzymatic browning, inhibition, 171
Equations
for attenuation factor, 269
for complex permittivity, 269
for electrical conductivity, 271
for energy, 269

for heating, 264, 270
for intrinsic impedance, 269
for penetration depth, 241, 270
for power absorption, 242, 263, 270
for relaxation time, 244, 254, 271
for specific heat of sucrose, 52
for wavelength in foods, 271
Equilibrium relative humidity. *See* Water activity
Esters, sucrose in, 279-280
Ethanol, as sucrose byproduct, 282
Ethers, sucrose in, 279
Evaporation, 19, 30, 33
equipment, 17, 23
Extra fine sugar, 37
Extruded cereals, 187
formula, 189

FASEB. *See* Federation of American Societies for Experimental Biology
Fat substitute, from sucrose 285
Fats
in human diet, 97-98
as enhancer of sweet taste, 75
use in candymaking, 123
FDA. *See* Food and Drug Administration
Federal Register, GRAS report on sucrose, 90, 98-99
Federation of American Societies for Experimental Biology
Select Committee on GRAS Substances, 89
Fenfluramine, as serotonin enhancer, 97
Fermentation
products, 281-283
sugar role in, 134-135, 165, 277
Fillings
for bakery foods, 145-146, 149-150
formulas and procedure, 149-150
for pies, 174-175
Filtration, 17, 32. *See also* Purification
Fine sugar. *See* Extra fine sugar
Firming, sugar and, 137
Flaking, in cereal manufacture, 185
Flavor differential, 40
Flavor enhancement by sugar, 67-68
in bakery foods, 132, 136, 140, 143-144
in beverages, 209-210
in cereals, 187-188
in dairy products, 153, 164
in processed foods, 165, 171, 178
Flavor grading of preserves, 219-220
Flavor perception, 68-69
Flavored syrup, inversion in, 205
Flavorings, sugar in, 160-163
Floc, analytical method for, 296-297
Flow diagrams
of beet sugar process, 31
of raw sugar mill, 18
of cane sugar refining process, 25
of yogurt production, 162
Foam-type cakes, 142-143
formulas, 143
Foam-type icings, 146
Foaming agents, 143
eggs as, 140
Fondant, 110-114, 128
formula, 111-112
in icings, 131
production, 106-107, 111
Food and Drug Administration
standards for preserves and jellies, 214-217
studies, 89
Task Force on Sugars, 90-92, 94-95, 98-99
Food dielectric. *See* Dielectric properties
Foods, in microwave cooking, 229-236, 249
Frappé
as aerating ingredient, 112
formula, 113

Free-flowing brown sugars, 40
Freezing
  excess sugar and, 160-161
  in fruit processing, 171
  high fructose corn syrup and, 153
  in ice cream making, 154-155
Freezing point, of sucrose solution, 53
Frequencies, microwave, 230-235, 250-251
Frosted nuts, microwave recipe, 262
Frostings. *See also* Icings
  for cereal, 185, 187, 190, 192
  microwave recipe, 259
Frozen dairy products, sugar role in, 152-155
Frozen novelties, 156
Fructose, in baking, 135
Fructose metabolism, 87
  glucose tolerance and, 92
Fruit butters and fillings. *See* Preserves
Fruit flavorings, 160
  for beverage powders, 173, 210
  citrus pureés as, 174
Fruit granulated sugar, 37
Fruit processing, sucrose in, 171
Fudge, 127
  formula, 128
  microwave recipes, 261
Functional purposes, of sugar, 165-181

Gasoline production, sugar role in, 282
Gelatin
  in cereal frostings, 192
  dessert mix, 176
  in icings, 149
Gelling agents, 175
Generally Recognized as Safe. *See* GRAS
Geography, of sugar sources, 1-2, 26-27
Geometry of foods, in microwave cooking, 251-252
Gingersnaps, surface cracking of, 146
Glazes. *See* Frostings
Glossary, 303-311
Glossing, invert sugar in, 145
Glucagon role, in glucose metabolism, 88
Glucose
  tolerance, 91-93
  transport in human metabolism, 85-88
Gluten, crumb texture from, 137
Glycemic response, 91
Golden brown sugar. *See* Light brown sugar
Grain size of sugar
  in beverage mixes, 210
  in candymaking, 106, 108
Graining, 106-110
Granulated sugar
  in baking, 130-131
  in gelatin desserts, 176
Granulated sugar products, 36-39
  analysis of, 42
  icings as, 145
Granulating, in cereal manufacture, 185
Grape jelly, formula, 225
Grapefruit pureé, formula, 174
GRAS status
  of sucrose, 89
  of sugars, 98-99
Grist, analytical method for, 291

Ham, sugar-cured, 180
Hand-rolled creams. *See* Plastic creams
Hard candies. *See* High-cooked candies
Heat. *See* Temperature
Heat of solution, of sucrose, 52-53
Heating. *See also* Temperature
  equations, 264, 270

microwave, 235
HFCS. *See* High fructose corn syrup
High-cooked candies, 116-122
formula and procedure, 119
microwave recipes, 261-262
High fructose corn syrup
in beverages, 198-199
in ice cream, 153
in preserves, 216
Homogenization, in ice cream making, 154
Honey, 48
Household microwave ovens, 234, 242-243
Hydrochloric acid, in invert sugar production, 109
Hydrolysis, of sugar, 135, 167
Hyperactivity, sugar consumption and, 95
Hyperglycemia, in diabetes mellitus, 88
Hypoglycemia, in diabetes mellitus, 88

Ice cream
contents, 154-155
formulas, 154
sugar in, 152
Ice milk, 155
formulas, 156
Iced tea mix, 177, 210
Ices, 157
Icings
formulas and procedure, 147-148
sugar in, 145
ICUMSA. *See* International Commission for Uniform Methods of Sugar Analysis
Icy textures, 153
Impurities, removal. *See* Clarification; Filtration
Individual preference, of flavor, 69, 76-77
Industrial cooking, microwave, 228, 234
Industrial sugar deliveries, 44
Insecticides. *See* Pest management
Institutional uses, of microwave power, 233-234
Insulin role, in glucose metabolism, 88, 92
International Color Units, 38, 40
International Commission for Uniform Methods of Sugar Analysis, 37-38, 51, 201-202
International Sugar Research Foundation, 277-279
International Union of Pure and Applied Chemistry nomenclature, 47
Intestines
fructose metabolism in, 87
sucrase activity in, 85
Intrinsic impedance
equation for, 269
microwave, 240
Invert sugar
as agglomeration agent, 170
in baking, 131
in beverages, 198
as crystallization retardant, 40
effect on graining, 108, 110, 116-117
in icings, 145
liquid, 39-40
naturally occurring, 48
production, 109, 112
reaction, 48, 200-207, 209, 216-217, 278
total, 40
Invertase
in baking, 135
in candymaking, 112
IUPAC. *See* International Union of Pure and Applied Chemistry

Jams. *See* Preserves
Jellies. *See also* Preserves
consistency, 219-220
formulas, 223-226

standards, 215-217
sugar role in, 212-227

Laboratory equipment, jelly-making, 222
Lemon-flavored iced tea mix, formula, 177
Lemon pie fillings, 174-175
formula, 174
Leuconostoc bacteria, reaction with sugar, 58
Light brown sugar, 40
Lipids. *See* Blood lipids
Lipoprotein transport, 94
Liqueurs, sugar in, 210-211
Liquid ferment dough, 132-133
Liquid invert sugar, for carbonated beverages, 39
Liquid sugars, 39-40
analysis, 43
properties, 59
Literature reviews, 277-278
Liver role, in glucose metabolism, 86-87
Livestock feed
sugar beet as, 27

Maillard reaction, 57, 125, 136-137, 166-167, 188
Mannose, in sugar metabolism, 87
Marmalade, 212. *See also* Jellies; Preserves
Marshmallow icings, formulas and procedure, 148
Mazzetta. *See* Frappé
Measurement of sweetness
psychometric techniques for, 72-73
validity of, 73-74
Mechanical mixing
in baking, 141
of dry mixes, 170
Medicine, sugar role in, 3, 212, 277, 283-284
Melanoidin formation, 137, 167
Metabolism, of sugars, 85-88, 90-97
Microbiological action
stability with sugar, 210
in sugar contamination, 61-62
Microwave cooking
load factor curve in, 263
major applications of, 256-257
oven power for, 263
problems, 255
recipes, 257-262
sugar role in, 228-275
time, 264
Microwave energy, 230-235, 242-248
equation for, 271
frequencies, 230, 242-248
Milk products. *See* Dairy products
Mineral fortification, sugar and, 166
Mints, formula and procedure, 116-119
Mixing methods, in baking, 141-142
Moisture
analytical method for, 290
in brown sugars, 40
in cakes, 140
in granulated sugar production, 38
in microwave cooking, 248
retention by icings, 145
Molasses
in baking, 131
citric acid from, 282
from sugar processing, 20, 24, 33, 41, 281-282
flavor of, 69
literature review, 278
properties, 59
Molasses kiss, formula and procedure, 123
Mousse, microwave recipe, 259
Mouthfeel
of bakery products, 145
of beverages, 209
of cereals, 188
of confectionery products, 106-110

National Food Processors Association, 39
National Institutes of Dental Research, 91
National Soft Drink Association, 208
"No-bake" pie fillings, 175
No-time breadmaking. *See* Short-time breadmaking
Nonindustrial sugar deliveries, 44
Nougat, 123
formula and procedure, 124

Oat flakes, formula, 189
Obesity. *See* Weight control
Odor, sweetness and, 75
Opaquing agent, titanium oxide as, 175
Optical rotation, of sucrose, 50-51
Osmotic pressure, of sucrose solution, 55-56
Ovens, microwave, 233-234, 242-243

Pasteurization, 153-154
microwave, 228
of preserves and jellies, 221
Peanut brittle, formula and procedure, 121-122
Pecan pie, microwave recipe, 260
Pectin
degradation inhibition, 171-172, 222
as gelling agent, 218
Penetration depth
equation for, 241, 270
microwave, 241, 265
Penuche, microwave recipe, 260
Peppermint divinity, microwave recipe, 260
Pest management, 12, 16-17, 28, 30
pH, 57, 109, 143, 200-201. *See also* Acidity; Alkalinity
Pharmaceuticals. *See* Medicines
Photosynthesis
sugar production in, 82-85
Physical constants, 268
Pie fillings. *See also* Pecan pie
formulas, 174-175
Plastic creams, 113-114
Plastics, sucrose in, 278
Polarization, analytical method for, 300-301
Polyester, from sucrose, 285
Post-ingestive effects, taste and, 77-78
Postprandial concentration, of glucose, 88
Power absorption, equations, 240, 263, 270
Powdered beverage mixes. *See* Dry beverage mixes
Powdered confectioners sugar, 39
in baking, 131
as flavor carrier, 69
in icings, 145
properties, 59
storage and handling, 62
Preservatives
as additive, 216
sugar as, 177-178
Preserves
consistency, 219-220
formula, 225-226
standards, 214-217
sugar role in, 212-227
Processed foods
meats as, 177-180
sugar role in, 165-181
Product uniformity, in quality control, 220
Properties, of sugar, 46-65, 103-110
Psychometric techniques, for sweetness measurement, 72-73
Publications, on sugar and health, 97-100
Puffed cereals, formulas, 190
Puffing, in cereal manufacture, 186-187

Purification. *See also* Filtration
  in sugar processing, 24, 32

Raisins, sugar coating for, 191-192
Rate of heating, equation, 270
Ready-to-eat cereals, sugar role in, 182-197
Refined sugar producers, 312-314
Refining process, 22-24, 32-33
  flow diagram, 25
Refractive index, of sucrose, 49-50
Refractometers, 50
  calibration, 205, 207
  in-line, 221
Refrigeration, in ice cream making, 154
Relaxation time, equation, 244, 254, 271
Research, sugar uses in, 278-285
Rotary-drum vacuum filters, 17
Rotary molded cookies, 144
R.T.E. cereals. *See* Ready-to-eat cereals

Saccharomyces cerevisiae. *See* Bakers yeast
Safety, of sweeteners, 98-100
Sanding granulated sugar, 37, 191
Sandwich cookie filling, formula and procedure, 149-150
Saponin, analytical method for, 298-299
Sausages, processes for, 178-179
Seasonal products, 159
Sediment, analytical method for, 292
Seedcane planting, 12-13
Seeding, 106, 115
Serotonin, in carbohydrate consumption, 96
Shape of foods, in microwave cooking, 251-252
Shelf-life extender
  corn syrup as, 119-120
  sugar as, 132, 165, 176, 179, 195, 210
Short-time breadmaking, 132-134
Shortening
  in cakes, 140-142
  in cookies, 143-144
  formulas, 140
  in icings, 147
Shredding, in cereal manufacture, 185-186
Shrinkage, of invert solutions, 202-203
Single-stage mixing, in baking, 141
Size
  of food mass, 252
  of sugar grain, 37
Snack cakes, 142
  formula, 143
Sodium bicarbonate, in invert sugar production, 109
Soft brown sugar, 40
Solids
  formation in sucrose inversion, 201-202
  ranges in beverages, 208
  ranges in preserves and jellies, 222-223
Solubility, sucrose and, 48-49, 104-106, 165-166, 170, 277
Solventless process, for sucrose, 280
Specialty sugars, 41
Specific gravity, of sucrose, 49
Specific heat
  in microwave cooking, 253-254, 266
  of sucrose, 52
Specks, analytical method for, 302
Spectroscopy, in flavor analysis, 69
Spice mixes, in processed meats, 178-179
Sponge cakes, 142
  formula, 143
Sponge dough breadmaking, 132-133

Stabilizers, 166
in gelatin desserts, 176
in ice cream, 153
in icings, 146
Standards
for bottlers, 208
of identity purity for sucrose, 89
for jams and jellies, 214, 217
Starch, swelling of, 142
Steam processing, 22, 33
agglomeration and, 170
in candymaking, 125-126
Steffen Process, 6
Sterilization, microwave, 228
Storage, 20-21, 29-30
of brown sugars, 40
of granulated sugar products, 60-61
of liquid sugars, 61-62
Sucrase, in sugar metabolism, 85
Sucrose. *See also* Sugar
as agglomerating agent, 169-170
biological properties, 58-59
chemical properties, 56-57, 82, 276-277
chemical reactions, 125, 278-281
chemistry, 46-48, 82
content of cereals, 192-193
crystalline, 36-39
in dairy products, 152-164
flavor, 66-69
glucose tolerance and, 92-93
inversion, 200-207
liquid, 39-40
nonfood uses of, 276-287
physical properties, 48-56, 64, 68, 82, 104-106, 277
research, 275-285
spun, 192
water activity and, 55, 165, 168-169
Sugar. *See also* Beet sugar, Cane sugar, Sucrose
analytical methods for white, 288-302
in beverage industry, 199
in browning, 136
on cereals, 195
chemical structure of, 83
concentration in baking, 142-143
consumption, 1, 7-8, 132
cultivation, 11-20, 26, 28-29
dust, hazards of, 61
flavor in foods, 66-70, 136
geography, 1-2, 26-27
health and, 88-100
history, 1-7, 22-23
human body and, 82-100
hydrolysis in baking, 135
in microwave cooking, 228-275
origin of name of, 2
from photosynthesis, 82-85
processing, 11-14, 17-21, 23-25, 27, 30-35
producers, 312-316
properties, 46-65, 103-110
storage and handling, 60-63
Sugar beets. *See* Beet sugar
Sugar industry, 1, 6-9, 26-27
affiliates, 316-318
producers, 312-316
terminology, 303-311
Sugar mill, flow diagram, 18
Sugar products, 36-45
deliveries, 44
Sugar Research Foundation. *See* International Sugar Research Foundation
Sugar uses
as chemical raw material, 276-281
in cosmetics, 284
in fermentation, 58, 277, 281-283
in medicines, 3, 212, 277, 283-284
in plastics, 278
Sugarcane. *See* Cane sugar; *see also* Seedcane
Sulfite, analytical method for, 294-295
Sunkist Growers Ridgelimiter, 222

Surface coatings. *See also* Frostings, Icings
  of foods, 191-192
  industrial, 279, 283-284
Surface cracking, of cookies, 146
Surface tension, of sucrose solution, 54
Surfactants, sucrose in, 279
Surgeon General's Report, 98
Sweet doughs, formulas and procedures, 138-139
Sweetened condensed milk, 158-159
  in ice cream, 153
  procedure, 159
Sweeteners
  in beverages, 198-211
  comparison of, 65
  consumption of, 7-8
  in ice cream, 155, 163
  mixtures of, 75
  safety of, 98-100
Sweetness
  of ice cream, 156
  measurement of, 71-73
  of invert sugar, 201, 207
  sucrose as standard for, 67
  of sugar, 40, 71-78, 153
  variation in, 175

Taste interactions, sweetness and, 74-76, 209
Temperature. *See also* Cooling; Heating
  effect on browning, 137
  effect on graining, 106-108
  effect on sweet taste, 75-76
  in ice cream making, 153-154
  in microwave cooking, 249, 262, 272
  stability with sucrose, 210
  sucrose inversion and, 201
Teratogenicity studies, 89
Terminology, 303-311
Texas Agricultural Experiment Station, 14-15
Texture. *See also* Graining
  of bakery products, 137-138
  of cereals for coating, 192
  of confectionery products, 107, 111, 122
  enhancement by sugar, 132, 140, 188
  of gelatin desserts, 176
  in ice cream, 153, 155-156
  of meat products, 179
Thermal conductivity, in microwave cooking, 254
Thermal decomposition, of sucrose, 56-57
Thin film cookers, 121
Time
  effect on sucrose inversion, 201, 204-205
  in microwave cooking, 263-264
Titanium oxide, as opaquing agent, 175
Tomato catsup products, 171-172
  formula, 172
Tooth decay. *See* Dental caries
Total invert syrup, 40
Tricalcium phosphate, as anticaking agent, 177
Turbidity, analytical method for, 289

United States National Committee on Sugar Analysis, 288
United States Sugar Corporation Research Station, 14
USDA. *See* U.S. Department of Agriculture
U.S. Department of Agriculture
  Food and Nutrition Briefs, 97
  glucose tolerance studies, 92, 94
  preserves standards, 219-220
  Sugarcane Field Station and Laboratory, 14
U.S. Pharmacopeia, National Formulary, 39, 89

Vacuum cooking
  in candymaking, 120-121
  for cane sugar processing, 17, 19
  in jelly making, 221
Vanilla fudge, formula and procedure, 128
Vapor pressure, of sucrose solution, 54
Viscosity
  of foods, 254-255
  of sucrose, 51-52, 192
Vitamin fortification, sugar and, 166
Volume decrease, in sucrose inversion, 202-203

Water absorption, 167
Water activity
  in microwave cooking, 236
  sucrose and, 55, 165, 168-169
Water levels, in cookies, 143
Water-spray process, agglomeration and, 170
Wavelength, 230-235, 244, 246, 250-251
  equation, 271
Weight control
  fat consumption and, 97
  sugar consumption and, 96-98
  Surgeon General's Report and, 98
Whey, in ice cream, 153
Whipping aids, sugar, 143
Wines, sugar in, 58, 210-211
Wire-cut cookies, 144
Yeast fermentation, of sugars, 58, 132
Yeast-leavened bakery foods, 132-139
Yogurts
  formulas, 161
  production flow diagram, 162
  sugar in, 161